34.95

Coal Liquefaction and Gasification Technologies

Edward C. Mangold • Mario A. Muradaz
Robert P. Ouellette • Oscar G. Farah
METREK Division of the MITRE Corporation

Paul N. Cheremisinoff
New Jersey Institute of Technology

ANN ARBOR SCIENCE
PUBLISHERS INC / THE BUTTERWORTH GROUP

TP
352
.C637
1982

Copyright © 1982 by Ann Arbor Science Publishers, Inc.
230 Collingwood, P.O. Box 1425, Ann Arbor, Michigan 48106

Library of Congress Catalog Card Number 81-69550
ISBN 0-250-40494-X

Manufactured in the United States of America
All Rights Reserved

Butterworths, Ltd., Borough Green
Sevenoaks, Kent TN15 8PH, England

PREFACE

During the past decade large increases in petroleum prices have had profound effects on the economies of the world. An outgrowth has been evaluations on contributions that can be made by new technologies utilizing coal to energy programs. This book resulted from a study by the Metrek Division of the MITRE Corporation, sponsored by the French national utility, Electricité de France, to explore opportunities for: (1) technologies for coal utilization being developed, and (2) basic research and development on coal utilization technologies.

This study reviews the state-of-the-art of selected liquefaction and gasification processes that have been developed with support from the United States. Information in this study was obtained from sources in the public domain, with contractor project reports to the U.S. Department of Energy as a major source. Conference papers and technical journal articles on process development were another major source often used. These documented sources were supplemented by conversations with personnel from the companies that are the process developers. Much of the information that is crucial to the technical development of the processes is proprietary and not obtainable by the general public.

The processes contained in this review are:

1. liquefaction—Exxon Donor Solvent, H-Coal, Solvent-Refined Coal I and II, Mobile Gasoline Synthesis, Fischer-Tropsch Synthesis and Zinc Halide Hydrocracking; and
2. gasification—Slagging Lurgi, Texaco, Combustion Engineering, COGAS and Shell-Koppers.

Acknowledgment and thanks are made to Electricité de France for its support in this work.

Edward C. Mangold
Mario A. Muradaz
Robert P. Ouellette
Oscar G. Farah
Paul N. Cheremisinoff

MANGOLD **MURADAZ** **OUELLETTE**

Edward C. Mangold holds a PhD in physics from Oklahoma State University. He spent four years with the U.S. Environmental Protection Agency Enforcement Division performing field studies at several power plants and industrial facilities, where new instruments and measurement methods were investigated for use in operational field measurements of pollution. For the last three years at the MITRE Corporation, Dr. Mangold has worked to improve the management control systems used by the Department of Energy for environmental and safety problems in nuclear, petroleum and fossil energy technologies. A state-of-the-art assessment of coal-based technologies was performed for gasification, liquefaction, fluidized-bed combustion and combined-cycle power plants.

Mario A. Muradaz is a Research Associate with the Metrek Division of the MITRE Corporation. For the past three years he has worked in Technology Programs, assessing the technical and economic merits of new industrial processes and technologies for Electricité de France. Mr. Muradaz has an extensive background in chemical engineering and holds a degree in economics from the University of Maryland.

Robert P. Ouellette is Technical Director of the Environment Division of the MITRE Corporation. Dr. Ouellette has been associated with MITRE in varying capacities since 1969 and had been Associate Technical Director since 1974. Earlier, he was with TRW Systems, Hazelton Labs, Inc. and Massachusetts General Hospital. He graduated from the University of Montreal and received his PhD from the University of Ottawa. A member of the American Statistical Association, Biometrics

FARAH   **CHEREMISINOFF**

Society, Atomic Industrial Forum and the National Science Foundation Technical Advisory Panel on Hazardous Substances, Dr. Ouellette has published numerous technical papers and books on energy and the environment. He is co-author of the comprehensive *Electrotechnology* survey series published by Ann Arbor Science.

Oscar G. Farah is Manager of International Programs, Energy Resources and the Environment Division at Metrek, a division of the MITRE Corporation. Before joining MITRE, he was a research scientist at the Naval Research Laboratories and the Carlisle Barton Laboratory of Johns Hopkins University, and was a member of the technical staff at Bell Telephone Laboratories and the Western Electric Company. Dr. Farah obtained a BS in mechanical engineering from Purdue University and a PhD in electrical engineering from Johns Hopkins University. He has published a number of papers on world energy demand and supply, transmission, and conservation, and is co-author of *Ethylene: Basic Chemicals Feedstock Material,* published by Ann Arbor Science.

Paul N. Cheremisinoff is Associate Professor of Environmental Engineering at the New Jersey Institute of Technology. He is a consulting engineer and has been a consultant on environment/energy/resources projects for the MITRE Corporation. A recognized authority on pollution control and alternative energy technologies, he is author/editor of many publications, including several Ann Arbor Science handbooks on pollution and energy, such as *Pollution Engineering Practice Handbook, Carbon Adsorption Handbook, Environmental Impact Data Book, Industrial and Hazardous Wastes Impoundment,* and *Environmental Assessment & Impact Statement Handbook.*

CONTENTS

Section 1: Liquefaction

1. Introduction/Background 3
   Coal Utilization 3
   Process Summary 4
   Plant Economics as a Function of Size 6
   European Utilization of Gasification and Liquefaction
      Technologies 7
   Initial Cost Estimates on Large Engineering Projects .. 9
   Recommendations Concerning Process Selection 11

2. Major Liquefaction Processes 13
   Exxon Donor Solvent (EDS) 13
      Process Description 13
      Historical Development and Status 14
      Technical Evaluation 16
      Pilot-Plant Operations Description of the 1-ton/day
         Pilot Plant 31
      Operation of the 1-ton/day Pilot Plant 34
      Project Assessment 37
   H-Coal Process 37
      Process Description 37
      Technical Evaluation 40
      The 250- to 600-ton/day (227- to 545-metric ton/day)
         Pilot Plant 47
      Commercial H-Coal Liquefaction Plant 50
      Economic Analysis 58
      Assessment 60
   Solvent-Refined Coal 60

Process Description	60
Technical Evaluation	67
Pilot-Plant Operations	71
Commercial Plant Design	73
Assessment	77

3. Minor Liquefaction Processes 79
   Mobil Gasoline Synthesis 79
      Process Description 81
      Preliminary Design of a Commercial Plant 83
      Mobil Commercial Plant Design 85
      Comparison of Mobil and Fischer-Tropsch Processes 87
      Economic Analysis 89
      Project Assessment 93
   Fischer-Tropsch Synthesis 93
      Fischer-Tropsch Process Description 94
      Assessment 97
   Zinc Halide Hydrocracking Process 99
      Process Description 99
      Technical Evaluation 99
      Process Economics 102
      Assessment 102

4. Common Problems in Liquefaction 105
   Environmental and Health Problems 105
   Liquefaction Process Problems 107
      Mechanical Equipment 107
      Wastewater Treatment 107
      Solid Waste Disposal 108
      Atmospheric Emissions 109
      Liquefaction Bottoms 109
      Cost of Hydrotreating Coal Liquefaction Products 111
   Economic Analysis of Coal Liquefaction Processes 114

Section 2: Gasification

5. Introduction 121
   Gasification Processes 121
   Summary of Gasification Processes 122
   Gas Cleanup Systems 123
   Mathematical Modeling of Gasifier Performance 128

Drying	129
Devolatilization	129
Gasification	129
Combustion	129
Kinetic Equations	130
Steady-State Differential Equations	130
Comparison with Test Data	131

6. Major Gasification Processes 133
    Slagging Lurgi 133
        Process and Manufacture 133
        Pilot-Plant Operations 151
        Technical Evaluation 168
        Economic Analysis 175
        Environmental Evaluation 188
        Assessment 208
    Texaco Coal Gasification Process 209
        Process and Manufacture 209
        Pilot-Plant Operations 211
        Technical Evaluation 220
        Environmental Analysis 223
        Assessment 223
    Combustion Engineering 224
        Description and Schematic 224
        Technical Evaluation 227
        Economic Evaluation 234
        Environmental Analysis 238
        Assessment 239

7. Minor Gasification Processes 243
    Shell-Koppers 243
        Description and Schematic 243
        Economic Analysis 247
        Assessment 248
    COGAS 248
        Description and Schematic 248
        Economic Analysis 255
        Pilot-Plant Operations 258
        Assessment 258

Index 261

LIST OF FIGURES

2-1.	Exxon donor solvent process	14
2-2.	EDS commercial plant study design update. Illinois coal market flexibility sensitivity case, configuration of units	30
2-3.	H-Coal process	39
2-4.	H-Coal ebullated-bed reactor	41
2-5.	Commercial H-Coal complex block flow diagram	54
2-6.	SRC process flowsheet	61
2-7.	Flowsheet for Wilsonville SRC pilot plant	62
2-8.	Filtration flowsheet	64
2-9.	Simplified flowsheet for the SRC-II process	67
2-10.	Simplified flow diagram for the proposed SRC-II commercial plant	75
3-1.	Mobil methanol-to-gasoline synthesis	80
3-2.	Schematic of fixed-bed pilot plant	82
3-3.	Mobil methanol-to-gasoline process	84
3-4.	Fischer-Tropsch synthesis	96
3-5.	Zinc chloride catalyst process	100
6-1.	Conceptual slagging Lurgi pipeline gas plant	134
6-2.	Gasification	144
6-3.	Shift conversion	149
6-4.	Coal feed rate vs oxygen loading with and without feed or recycle solids-laden tar	158
6-5.	Effect of slag composition and steam/oxygen ratio on slag tapping	160
6-6.	Effect of coal swelling index and feed fines concentration on distributor torque	163
6-7.	Summary of problems during technical support program	166
6-8.	Gas cost variation with DCF rate of return	189
6-9.	Gas cost variation with coal cost	190

6-10. Gas cost variation with capital investment 191
6-11. Gas cost variation with operating cost 192
6-12. Gas cost variation with capital investment (utility financing economics) 193
6-13. Gas cost variation with coal cost (utility financing economics) 198
6-14. Gas cost variation with debt interest rate (utility financing economics) 199
6-15. Gas cost variation with maintenance and operating expense (utility financing economics) 200
6-16. Gas cost variation with by-product value (utility financing economics) 201
6-17. Gas cost variation with plant yield (utility financing economics) 202
6-18. Texaco gasifier 210
6-19. Simplified flowsheet of Texaco process at Montebello, CA, during the 1950s 214
6-20. Pilot-plant flowsheet 217
6-21. Air gasification for fuel gas manufacture 218
6-22. Burner efficiency vs burner ΔT 219
6-23. CO exhaust emission comparison. Burner rig data vs engine data 220
6-24. Combustion Engineering entrained-bed coal gasification process 226
6-25. Schematic flow diagram of the CE pilot plant 228
6-26. Material and flow schematic with process data for 4.5-Mg/hr pilot plant 229
6-27. Schematic of conventional steam plant with atmospheric gasifiers 235
6-28. Schematic of combined cycle plant with supplementary fired boiler and atmospheric gasifiers 236
6-29. Schematic of combined cycle plant with heat recovery boilers and atmospheric gasifiers 237
7-1. Schematic flow diagram of the Shell-Koppers process 244
7-2. Overall thermal efficiency 246
7-3. Cost of gasification product 247
7-4. COED process flow diagram 250
7-5. Pyrolysis—gasification 251
7-6. Gasification-combustion system 252
7-7. Illinois coal gasification group, COGAS commercial plant concept 253

LIST OF TABLES

1-1.	Liquefaction Process Data Summary	5
1-2.	Economic Analysis: Liquefaction Process	6
2-1.	ECLP Test Program: Summary of Selected Development Programs	18
2-2.	Schedule for 2.5-year ECL Test Program	19
2-3.	Analyses of Feed Coals Run in RCLU	22
2-4.	Highest Liquid Product Yields for EDS Program Coals at 1500 psig (RCLU Data)	23
2-5.	Exxon Donor Solvent Commercial Plan Study Design Update: Construction Cost Estimate	31
2-6.	Investment Expenditure Schedule	31
2-7.	Exxon Donor Solvent Commercial Plant Study Design Update: Annual Operating Expenses	32
2-8.	Exxon Donor Solvent Commercial Plant Study Design Update: Assumptions for Computing Annual Operating Expenses	32
2-9.	Exxon Donor Solvent Commercial Plant Study Design Update: Material and Product Balance	33
2-10.	Equipment Sizes for H-Coal and H-Oil Facilities	46
2-11.	Pilot-Plant Phase III Technical Management Plan	50
2-12.	Projected Pilot-Plant Yields: Boiler Mode	50
2-13.	Projected Pilot-Plant Yields: Intermediate Mode	51
2-14.	Analysis of Coal	51
2-15.	Typical H-Coal Process Yields, 3-ton/day Process Unit, Continuous Catalyst Replacement	52
2-16.	H-Coal Commercial Plant Design Material and Product Balance, Stream-Day Basis	52
2-17.	Overall Thermal Efficiency of Commercial H-Coal Plant	53
2-18.	Coals Tested for SRC-I	66
2-19.	Ultimate Analysis, Kentucky Coals No. 9 and 14, Colonial Mine	70

2-20.	Product Yields in SRC-I and SRC-II Modes (Kentucky Coals Nos. 9 and 14, Colonial Mine)	70
2-21.	Coal Feedstock Composition	73
2-22.	Yield Data, Powhatan Coal Case	74
2-23.	Input, Output and Efficiency (Stream-Day)	74
2-24.	Capital Costs	76
2-25.	Annual Operating Cost	77
3-1.	Wyoming Subbituminous Coal Properties	87
3-2.	Mobil Methanol-to-Gasoline Commercial Plant Mass Balance (Stream-Day Basis)	88
3-3.	Investment Cost Assumptions	89
3-4.	Investment Costs	90
3-5.	Operating Cost Assumptions	90
3-6.	Operating Costs	91
3-7.	Economic Analysis Assumptions	91
3-8.	Unit Costs	92
3-9.	Fischer-Tropsch Commercial Plant Mass Balance (Stream-Day Basis for Comparison with Table 3-2)	98
4-1.	Common Liquefaction Process Steps	110
4-2.	Cost of Upgrading Coal Liquids to Turbine Fuels	113
4-3.	Products and Plant Capital Cost	115
4-4.	Product Value Factors	115
4-5.	Process Comparison	116
5-1.	Characteristics of Gasification Processes	124
5-2.	Gasification Process Data Summary	125
5-3.	Gasification Process Economic Analysis	126
5-4.	Effects of Processing Roland Seam Coal in Three Different Reactors	131
6-1.	Commercial-Size Gasification Plant Inputs and Products	135
6-2.	SNG Properties	136
6-3.	Gasifier Operating Condition	138
6-4.	Alternative Coal Operating Conditions	140
6-5.	Material Balance: Gasification	141
6-6.	Stream Analysis: Gasification	142
6-7.	Material Balance: Shift Conversion	145
6-8.	Stream Analysis: Shift Conversion	146
6-9.	Heat Recovery	150
6-10.	Material Balance: Gas Cooling	150
6-11.	Stream Analysis: Rectisol	152
6-12.	Material Balance: Methanation	155
6-13.	Stream Analysis: Methanation	156

6-14. Material Balance: Sulfur Recovery 157
6-15. Slag Tapping Performance for Selected Runs 159
6-16. Effect of Tar Recycle on Product Gas Composition During TSP Run 13: Flare Gas Analysis During Periods of Tar Recycle to the Gasifier 161
6-17. Effect of Tar Recycle on Product Gas Composition During TSP Run 13: Flare Gas Analysis During Periods of No Tar Recycle to the Gasifier 162
6-18. Performance of Internal Components of the Pilot-Plant Gasifier 164
6-19. Summary of Rationalized Material Balances (Basis: One Ton of Moisture- and Ash-Free Coal or Coke Feed) 167
6-20. Summary of Rationalized Heat Balances (Basis: Coal Higher Heating Value = 100.00) 168
6-21. Component Scale-up and Risk Analysis 170
6-22. Summary of Program Costs 175
6-23. Overall Heat and Material Balance 176
6-24. Plant Costs Summary 180
6-25. Capital Investment 182
6-26. SNG Product Gas Cost Alternative Coal Comparison 183
6-27. Operating Cost 184
6-28. Raw Material Summary 184
6-29. Catalyst Summary 184
6-30. Chemical Summary 186
6-31. Labor and Benefits Summary 188
6-32. By-product Summary 188
6-33. Product SNG Selling Price ($/$10^6$ Btu): DCF Method 194
6-34. Total Capital Requirement (Utility Financing Economics) . 195
6-35. Total Maintenance Requirements, Utility Financing Economics 195
6-36. Annual Operating Costs (Utility Financing Economics) ... 196
6-37. SNG Cost of Production (Utility Financing Economics) ... 197
6-38. Process Atmospheric Emissions from the Commercial Plant 203
6-39. Plant Component Pollution Potential During Normal Operations 204
6-40. Gasifier Operating Parameters 212
6-41. Coal Analysis (Illinois No. 6) 213
6-42. Gasifier Performance 213
6-43. Typical Results for the Texaco Gasifier at Montebello, CA, During the 1950s 216
6-44. Combustion Chamber Parameters 219

6-45. Liquefaction Residue Evaluation Program 220
6-46. Summary of Coal Liquefaction Residue Evaluations Completed July 1977–June 1978 221
6-47. Pilot-Plant Data from Gasification of H-Coal Syncrude Vacuum Tower Bottoms 222
6-48. Gasifier Material and Energy Balances 232
6-49. Coal Used and Production Gas Composition 234
6-50. Range of Product Gas Compositions 237
6-51. Operating Cost Data 239
7-1. Shell-Koppers Gasifier Operating Conditions 245
7-2. Shell-Koppers Coal Gasification Dry Synthesis Gas Composition 245
7-3. Shell-Koppers Coal Gasification Coal Feed Analyses 245
7-4. Shell-Koppers Coal Gasification Relative Amounts of Coal, Oxygen and Steam for Different Coals 246
7-5. Summary of COED Process Development Unit: Results on All Coals Tested 249
7-6. Properties of Illinois No. 6 Coal 254
7-7. Product and By-product 254
7-8. COGAS: Composition and HHV of Synthetic Pipeline Gas . 255
7-9. COGAS: Fuel Oil Product Analyses 255
7-10. COGAS: Naphtha Product Analyses 256
7-11. COGAS: Composition of Light Hydrocarbons 256
7-12. COGAS: Commercial Plant Capital Requirement 256
7-13. COGAS: Annual Operating Cost for a Commercial Plant .. 257
7-14. Bases for First-Year Economic Analysis (Utility Financing) 257
7-15. First-Year Gas Price Calculations, Illinois No. 6 Seam Coal, Utility-Type Financing 257

SECTION 1

LIQUEFACTION

CHAPTER 1
INTRODUCTION/BACKGROUND

COAL UTILIZATION

The factors that affect the types of coal that have been tested as gasification raw material include:

- coal reserve abundance and geographical distribution;
- distribution of American population and labor force;
- coal cost and suitability for other uses;
- low coal price for a plant sited at a strip mine;
- transportation problems;
- requirement for sulfur removal if used in direct combustion;
- suitability of the coal to the gasification and liquefaction process conditions, economics and yields; and
- political and environmental factors.

Three ranks of coal are considered for gasification and liquefaction feedstock: bituminous, subbituminous and lignite. American bituminous coal reserves are concentrated in the eastern half of the country and often contain 2–5% sulfur content by weight. Very large subbituminous and lignite reserves exist in the remote and sparsely populated West. They often contain less than 1% sulfur, but can have high ash and oxygen content and other objectionable properties.

Although the largest coal reserves are located in the West, the railroad network is not adequate to transport large quantities to the East, where the majority of the population resides. Gasification and liquefaction of the Western reserves provides a method of utilizing the resource, as well as converting it to a more convenient and less polluting form. Although Western coal is less expensive, the construction of plants is more difficult because of lack of an experienced labor force in this remote area.

4   LIQUEFACTION

Politically, Eastern interests tend to promote synthetic fuels plants because that is the location of the present coal industry. Western states often discourage the development of plants because of the environmental impact produced by large mining and conversion facilities.

The present American fossil fuel utilization pattern is heavily oriented toward liquid transportation fuels and natural gas. Liquefaction plants are designed to produce naphtha and light fuel oil, rather than heavy utility boiler fuel. The gasification program has concentrated on producing synthetic pipeline gas and chemical feedstock, with industrial and utility fuel gas production receiving less emphasis.

PROCESS SUMMARY

A summary of the principal liquefaction process parameters has been assembled in Table 1-1. Although every attempt has been made to generate comparable data, differences between processes not adequately stated in documentation from the process developer can lead to discrepancies in comparing results.

A limited list of economic parameters has been assembled in Table 1-2. Again, it is not always possible to classify costs into the specific categories desired because of the lack of detail and lack of information available. The capital cost includes the engineering and erection costs, site preparation and contractor's fees. Interest charges, environmental studies and other total capital costs are not included. Startup costs involving initial test operations and catalysts and chemicals are included where the data were available. Most of the estimates were prepared in limited detail, only to determine the basic level of project feasibility, and were not prepared on a comparable basis.

The proposed plant designs use process conditions chosen by the developers to optimize facility cost and product throughput. Numerous other processing conditions have been investigated experimentally. The developer's choice of conditions that result in maximum profitability and optimum operating conditions is influenced by:

- the type of coal used as feedstock;
- the recent shift from boiler to transportation fuel regarded as the desirable products;
- the desire to minimize C_1 to C_4 light gas production;
- the degree of hydrotreatment required to reduce sulfur and nitrogen content in the product to environmentally acceptable levels;
- the minimization of expensive hydrogen production;

INTRODUCTION/BACKGROUND 5

Table 1-1. Liquefaction Process Data Summary

	Exxon Donor Solvent	H-Coal	SRC-II	Mobil Methanol	Fischer-Tropsch
Name of Coal	Illinois #6	Illinois #6	Powhatan	Wyodak	Wyodak
Type of Coal	Bituminous	Bituminous	Bituminous	Subbituminous	Subbituminous
Coal Composition	Table 2-3	Table 2-12	Table 2-19	Table 3-1	Table 3-1
Material Balance (Stream day basis)					
Coal in (tons)	30,000	20,000	35,845	26,334	27,792
Electric Power in (Excess) [MW(e)]	184	NA	10.4	(5.3)	(3.31)
Raw Water in (tons)	NA	29.7	NA	37,725	39,840
Synthetic Gas (10^6 scf)	0	3,510	51	148	173
Propane (bbl)	2,610	3,290	28,433	1,555	1,107
Butane (bbl)	2,460	15,260	2,976	2,205	146
Naphtha (bbl)	23,200	27,940	17,120	22,045	13,580
Fuel Oil (bbl)	36,970	570	56,303	0	2,930
Sulfur (tons)	996	119	1,181	61	61
Ammonia (tons)	132	Table 2-15	183	103	103
Phenol/Alcohol (bbl)	480				1,825
Energy Balance	NA	75	Table 2-21	Table 3-2	Table 3-9
Thermal Efficiency (%)	64		72	62	58

6 LIQUEFACTION

Table 1-2. Economic Analysis: Liquefaction Process

	Exxon Donor Solvent	H-Coal	SRC-II	Mobil Methanol	Fischer-Tropsch
Capital Cost ($ million)					
Construction	3,773	1,000 (est)	1560	1471	1608
Catalysts & Chemicals			20	6.0	3
Startup Costs					
Capacity Factor (%)	85	90	NA	NA	NA
Operating Life (yr)	30	20	NA	20	20
Construction Time (yr)	5	4	5	4	4
Annual Operating Cost ($ million)					
Coal	316	$6.74/bbl	322.0	62.1	64.3
Other Raw Materials		0.12	3.6	1.4	1.4
Catalysts and Chemicals	23	0.74	8.0	5.0	5.9
Labor and Benefits	82	0.66	13.4	36.4	43.4
Maintenance Supplies			31.8	24.7	24.7
Maintenance Labor	155	1.80	15.6	27.5	
Other		1.57	23.4	(6.2)	76.1
By-product Sales	(40)	(3.57)			(6.0)
Required Personnel					
Operation	1071	NA	305	251	359
Maintenance	317		129	NA	NA
Administrative	269		169	NA	NA

- the technology used to separate coal ash from the liquid product;
- the maximization of plant thermal efficiency; and
- the avoiding of operating problems such as plugging caused by high bottoms viscosity.

Because of differences in these fundamental parameters, as well as numerous differences in detailed design, comparisons between two different processes are often misleading. Only engineers with a detailed knowledge of the plants and processing variations that are possible can make a meaningful judgment. For an additional description of these problems consult the Engineering Societies Commission on Energy study (summarized in Chapter 4).

PLANT ECONOMICS AS A FUNCTION OF SIZE

The budgetary estimates that have been performed are tentative because the plants of commercial scale are only conceptual. The first design of commercial-scale equipment will not be completed until the end of 1980. Although synthetic fuels technologies have been under development for decades, the economics have always been less attractive than competing fuels from petroleum. Until recent petroleum price increases were made, there has been little incentive to move ahead at a pace exceeding that required for an orderly research and development program. Until government policies are firmly established and financial incentives are provided, all cost and schedule estimates for commercial scale operations will remain indefinite.

The results of the economic analyses that have been performed indicate that the costs of capital recovery and high interest rates predominate. Next in order of magnitude is usually the cost of the coal. Labor and maintenance costs and utilities are the smallest contributors.

Detailed cost estimates are often only available for the very large commercial plants that have been defined as being appropriate for American needs. Economy of scale is achieved by utilizing equipment that is as large as scale-up laws, costs and problems of equipment fabrication, and transportation of the completed pressure vessels allow. The plants are complete self-sufficient facilities consuming in excess of 20,000 ton/day (18,140 metric ton/day). Together with their mines and waste disposal areas, they dominate surrounding land use and have strong economic effects on the adjacent communities. Since these technologies may be used at a much smaller scale in another country, with different land usage requirements and different environmental regulations, it is difficult to scale process costs to a smaller size unless there is a much better definition of the expected conditions and regulations.

EUROPEAN UTILIZATION OF GASIFICATION AND LIQUEFACTION TECHNOLOGIES

In evaluating a synthetic fuel technology for European utilization, the price and quality of the available coal must be considered with the application. An inexpensive, low-quality coal may be suitable for gasification, but not for direct liquefaction or conventional combustion. A high-quality bituminous coal having low sulfur and ash may have such a large demand for other uses that the high price makes the economics of liquefaction unattractive.

The applications of synthetic fuels are diverse and include:

- synthetic pipeline gas;
- light hydrocarbon gas petrochemical feedstock;
- gasoline for motor fuel;

8 LIQUEFACTION

- intermediate distillate fuels;
- heavy boiler fuel oils or solids;
- chemicals, including benzene, xylene, toluene and phenols;
- intermediate- and low-Btu gaseous fuels for boilers and processes;
- fuel gas for specialized applications such as combined cycle power plants; and
- chemical synthesis gas for metallurgical ore reduction, refinery processes, ammonia and methanol synthesis and hydrogen production.

All coal liquefaction processes produce a spectrum of products, although some are more specifically directed toward a single product. A higher price may be the penalty paid for controlling the product that is produced. Gasification processes, however, are specifically engineered for the specific intended application. The design of a commercial plant and economic analysis of a process may be performed specifically to fit the application. Liquefaction processes may operate 10 years from now in a market structure with prices and market segments served that are considerably different from historical patterns. If this occurs, the present concepts of optimum plant design could change significantly.

The initial commercial utilization of synthetic fuels technology can fulfill political, economic or strategic objectives, which each country must define. Dependability of a source of supply during a time of national crisis has historically been the greatest motivating factor for synthetic fuels production. Dependence of national policy on foreign sources of petroleum imports is a current political and strategic concern. From the purely economic viewpoint, petroleum-based fuels for three decades have been, and still are in most instances, the source of the lowest priced products. Alterations in this pattern may occur rapidly. Balance of payments considerations and stimulation of the local economy encourage the utilization of coal. Finally, public opinion may demand that action be taken and spur the utilization of technologies where other incentives are not adequate.

The available applications must be considered for individual installations and potential for wide-scale utilization. Synthetic fuels processes can be applied to activities as diverse as electric power generation, automotive engine fuels, heating of small buildings, industrial process heat and chemical feedstocks. A corresponding diversity exists in the size of facility to be built. The smallest facilities can serve one industrial plant with minimum investment and local impact. The use of coal for electric power generation and liquid fuel production will produce major sized projects dominating the economy of the local region.

The selection of an optimum gasification or liquefaction process also depends on the application for which the unit is intended. Most of the U.S. development effort in gasification has been expended on processes to produce synthetic pipeline gas because of the large potential market. Optimum processes have high thermal efficiency, high initial methane production and can tolerate features such as tar and oil production and slow response to changing demand. Gasification processes for combined cycle power plants, in contrast, must feature rapid variability of output for load following, controllability during process upset and ease of integration with the gas turbine and steam cycles. Lack of tar and oil production is an advantage since it simplifies operations. If a gasifier is to be used as a chemical feedstock producer, tars may or may not be desirable, depending on the application. Selection of the optimum process may also depend on the H_2/CO ratio that is desired. The final application influencing process selection is process heat and boiler fuel. Retrofitting existing heaters and boilers may require higher-heat-content gas to maintain capacity.

INITIAL COST ESTIMATES ON LARGE ENGINEERING PROJECTS

The record that has been established in the United States for cost estimates on large projects has been studied by Merrow at the Rand Corporation. Merrow has observed that on a large number of occasions initial project estimates have been low by a factor of 2 to 5. The reasons for these large overruns have been investigated, and the procedures by which cost estimates are made on large engineering projects are reviewed. Some of the study findings and conclusions that are important to selection of coal gasification and liquefaction technologies follow.

1. Project cost estimates are almost always low.
2. The earlier a cost estimate is made in the history of a project, the lower it is likely to be.
3. Cost estimates on commonly conducted activities, such as over 160 buildings, highway construction and river and harbor projects studied, usually only increase 25–60% over the estimated cost.
4. First-of-a-kind or one-of-a-kind projects incur larger cost overruns than projects with which the builders are familiar. For the 15 publicly funded projects studied, the average overrun was greater than 100%.
5. Project management quality has a significant effect on the size of a cost overrun.

10   LIQUEFACTION

6. The largest factor responsible for large cost overruns is changes in the project scope and design.
7. The accuracy of cost estimates increases as the project design definition approaches completion.
8. The larger the scale-up factor over existing demonstrated technology, the greater the risk of a large cost overrun.

Estimates of cost performed early in the evolution of a project are frequently low for several reasons. There is an inherent bias on the part of technical personnel and political backers to make the undertaking appear as attractive as possible. Outside factors such as inflation rates and the real increase in the prices of equipment and materials are difficult to estimate. However, the most important reason for low initial estimates is that it is not possible for the estimators to define the project scope and technical complexity in detail. This is particularly true where a one-of-a-kind or first-of-a-kind is attempted. It also applies strongly to projects that are stretched out over a long time and subject to considerable political, legal and regulatory influences. Whether changes are made for technical reasons or to alter the size of the effort, changes almost always increase costs. The earlier in the evolution of the project design the cost estimate is made, the larger the number of unforeseen problems and unappreciated details.

Since delay adds to costs, a prompt and efficient decision-making apparatus is an effective cost controller. However, where significant governmental input controls a project, a lengthy decision-making system and diffuse, imprecise decisions are more likely to be found. Environmental regulation changes and legal obstructions will add to the scope of an effort and require additional time for design and construction. Many ancillary project costs, such as road and parking lot construction, utility connections and other incidentals, are not included in early estimates but can ultimately add 50–100% to the costs of the basic elements that are the subject of attention during the early phase.

Merrow defines four types of cost estimates made during a project. Initial estimates are made before a definitive statement of project scope and technical process. A preliminary cost estimate is made after 5–10% of the design is completed. The definitive estimate, to which fabrication and construction contracts are written, is not made until the design is 90–100% completed. Since construction often starts on projects before completion of the design, the definitive estimate may be moved to an earlier step in the sequence than that which produces the highest accuracy. Because more details are visible to the estimator, and the effects

INTRODUCTION/BACKGROUND 11

of changes in the project design have been absorbed, the later estimates are higher in price and are more accurate.

RECOMMENDATIONS CONCERNING PROCESS SELECTION

Application of this experience to selection of a new coal gasification or liquefaction technology indicates that a premature selection should be avoided. Most of the technologies being developed by the United States are a factor of 100 or more below the commercial plant size, as defined for American needs. Based on past experience, it can be predicted that there is a high probability that scale-up will encounter technical problems, develop process improvements and be influenced by intangible factors, such as environmental opposition, vested political interests, international developments, and legal and regulatory challenges.

Using data from the present pilot-plant scale of operation, it is highly unlikely that an economic evaluation of a synthetic fuels process vis-a-vis competing petroleum or renewable fuels will be accurate for the time period in which commercial operations would take place. Technical improvements and opportunities for process cost reduction can only be incorporated over a long evolutionary development period. Premature commitment to a project construction with a large scale-up in size has in past experience lead to cost overruns of several hundred percent.

The three American direct liquefaction processes are in a critical stage of development. The H-Coal and Exxon Donor Solvent processes will have 250-ton/day (226-metric-ton/day) pilot plants operating in 1980. The Solvent Refined Coal I and II processes will have the detailed designs of 6000-ton/day (5424-metric-ton/day) demonstration scale plants completed by the end of the year. These plants will be more than 100 times larger than the existing 50-ton/day (45.3-metric-ton/day) pilot plant that has completed three years of operation. By 1982 a clearer selection can be made on the direct liquefaction process technical feasibility and economics, based on the two pilot-plant results and the completed demonstration plant design. Yields will be better established, operating and maintenance requirements defined, and costs estimated more accurately.

Many individuals in the United States have attempted to analyze cost differences between synthetic fuels processes. All have reported difficulties in achieving comparisons that were not subject to serious question.

When different organizations make estimates they use different procedures based on their experience and style of operation. The basic assumptions on which the process was designed and costs estimated are often not available, or are not comparable. Estimates have been made when processes were in different stages of development, and in different years. Many errors and gross presumptions have been found. The experience and skill of the estimator is impossible to determine, as is the degree of care and concern exercised. In many early estimates, the project team may concentrate on the basic processes of the project and neglect the requirements and costs of auxiliary facilities such as wastewater treatment.

Unless estimates of two competing processes are made by the same organization, expressly using the same method of estimating, the uncertainties involved will probably be larger than the accuracy of the estimate. Differential comparisons, such as the Mobil® gasoline process vs Fischer-Tropsch synthesis, or different types of gasifiers used in combined cycle power plants, are more accurate than absolute price prediction.

BIBLIOGRAPHY

Merrow, E. S., S. W. Chapel and C. Worthing. "A Review of Cost Estimation in New Technologies: Implications for Energy Process Plants," Report R-2481-DOE, Rand Corporation (1979).

Popper, H., Ed. *Modern Cost Engineering Techniques* (New York: McGraw-Hill Company, 1970).

CHAPTER 2
MAJOR LIQUEFACTION PROCESSES

EXXON DONOR SOLVENT (EDS)

Process Description

The EDS process sequence (Figure 2-1) is designed to maximize the production of liquid products. Feed coal is crushed, dried and slurried with the hydrogen donor solvent. The role of the solvent is to disperse the coal and transport it through the liquefaction system, and to donate hydrogen to the coal to promote liquefaction. The slurry is heated, mixed with hydrogen, and fed into a simple upward plug flow reactor, where it is maintained at a fixed temperature for a given residence time. The reactor effluent is separated by a series of distillation steps into the donor solvent (which is recycled), light hydrocarbon gases, naphtha, and medium and heavy distillate oils. A heavy vacuum bottoms stream, containing liquids with a boiling point exceeding 1000°F (538°C), unconverted coal and coal mineral matter, contains half of the weight of the feed coal. Recycle solvent is hydrogenated in a fixed-bed catalytic reactor with conventional petroleum technology using commercially available catalysts. Separation of solvent hydrogenation from coal processing is the unique feature of the Exxon process. It prolongs catalyst lifetime by treating a distillate oil which avoids contact with the catalyst-poisoning contaminants in the coal.

The heavy vacuum bottoms stream is processed to produce additional liquids and a low-Btu fuel gas for plant operations. An Exxon commercial petroleum process, Flexicoking, is being adapted to process residual liquids from coal liquefaction. All of the organic material in the bottoms is recovered as liquid product or combustible gases. Hydrogen for the process can be produced by steam reforming of light hydrocarbons produced in the liquefaction process, partial oxidation of vacuum distillation tower bottoms, or raw coal.

14 LIQUEFACTION

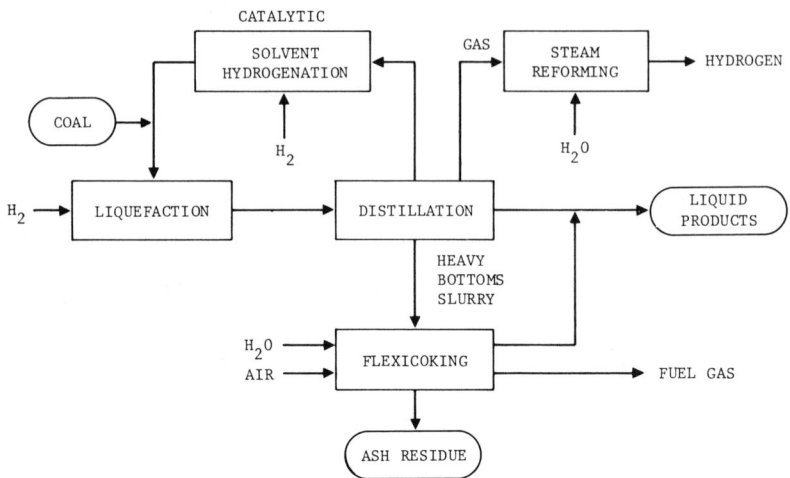

Figure 2-1. Exxon donor solvent process (source: "Development of the EDS Coal Liquefaction Process").

Historical Development and Status

The EDS project was begun by Exxon Research and Engineering Company in 1966. In 1976 additional participants were introduced into a 6.5-year development program that is 50% funded by the U.S. Department of Energy (DOE) and 50% by the Carter Oil Company, Phillips Petroleum Company, Atlantic Richfield Company, the Japan Coal Liquefaction Development Company and the Electric Power Research Institute. The total cost of the 6.5-year program is expected to be $240 million, with the 250-ton/day (227-metric ton/day) pilot plant costing $113 million to construct. The primary goal of the program is to achieve a state of commercial readiness by 1982. This means that the technology should be available at the end of the project to design and build a full-scale pioneer commercial plant that could begin operation before 1990, with a reasonable and acceptable level of risk.

The status of the program is that process development has been conducted on several small, bench-scale and pilot-plant units, and a 250-ton/day pilot plant is under construction at Baytown, TX. Initial shakedown activities began in the summer of 1979 as sections of the plant were completed. Coal processing operations will begin in April 1980 as the initial liquid production from three types of coal. The operations of the 250-ton/day pilot plant will demonstrate the operability of the liquefaction section and will obtain the data required to scale up the size of the equipment by a factor of 100 to that required for a 25,000-ton/day (22,700-metric-ton/day) commercial facility. Key objectives are demonstration of unit operability, acquisition of design data, and confirmation of product quality and yield. Demonstration of sustained operation at suitable solvent quality and satisfactory operation of the pumps and valves used to move the slurry are also key objectives. Scale-up data for commercial facility design will be obtained from the slurry preheat furnaces, liquefaction reactors, slurry dryers and vacuum fractionation unit. Data on product yields and quality will be checked against previous process studies, and the products generated will be used in product tests for commercial utilization. During the 2.5 years of pilot plant operations, an architect/engineering design of the pioneer commercial plant will also be performed so that at the end of the pilot plant operation, the design base and scale-up information required to build the commercial facility will be available. If the economic outlook is favorable or government support is obtained, detailed design of a commercial facility will then begin.

Future modifications to the EDS program may include process improvements that are under study by Exxon and other project sponsors. Currently active process improvement studies are directed toward increasing product yields and improving process operability and efficiency. For example, one method for increasing the liquid yields for certain coals is by recycling coal liquefaction bottoms. The yields of liquids from Illinois coal can be increased from 34 to 45% of dry feed coal weight by recycling the heavy bottoms through the liquefaction reactor to increase the residence time. Longer residence time without recycle does not lead to the same liquid yield increase because bottoms conversions to liquids is offset by hydrocracking of light liquids to gas. When the same processing technique was applied to Wyoming coal, insignificant yield increases were observed at the standard solvent-to-coal ratio. The liquid yield of Wyoming coal was increased by 10% when the solvent-to-coal ratio was increased by 50%, because of the increased donor hydrogen capability. Increasing the solvent-to-coal ratio produces a correspondingly larger stream of spent donor solvent and increases the size of the facilities required to hydrogenate the solvent.

To use the higher liquid yields, the overall thermal efficiency of the process must be increased to maintain an energy-balanced plant. Using figures from a 1976 study design in which the Illinois base case design produced 43% net liquids and a thermal efficiency of about 63%, the analysis shows that a liquid yield of 50% could only be obtained in an energy-balanced plant if the thermal efficiency were increased to 70%. Ideas for increasing liquid yield and achieving higher thermal efficiency

16 LIQUEFACTION

are being incorporated in the engineering program and are being included in design studies.

The optimum technique for processing bottoms will be investigated through studies of Flexicoking, and the Texaco and Shell-Koppers partial oxidation processes for producing hydrogen.

Technical Evaluation

Historical Development

The predevelopment phase occurred between 1966 and 1973, when the basic process for coal liquefaction was investigated. The present configuration and process flow scheme were selected from general options on the basis of research performed in experimental equipment ranging in complexity from 100-cm^3 batch units to a continuous 0.5-ton/day (0.45-metric-ton/day) integrated pilot plant. The reactor configurations that were evaluated included stirred tank, recirculated tubular, ebullating bed and tubular plug flow.

Several recycle solvents were evaluated on bituminous and subbituminous coals, with and without hydrogenation. Ranges of temperatures, pressures and residence times were evaluated in both the liquefaction and solvent hydrotreating reactors. Based on the results, a hydrogenated recycle process was selected for further development using noncatalytic, tubular plug flow liquefaction reactors that eliminated the need for ash tolerant catalysts. Vacuum distillation was identified as the most promising product separation technique. Engineering studies evaluated the laboratory data to develop conceptual commercial plant designs, study processing alternatives and estimate costs.

During the second phase, in 1974 and 1975, the basic design of the 250-ton/day coal liquefaction pilot plant was established. A 1-ton/day pilot plant was constructed and initially operated. A supporting laboratory and engineering research and development (R&D) program more closely defined operating conditions for the selected process. These studies are continuing as the 250-ton/day pilot plant nears completion.

The 1-ton/day pilot plant began operation during the first quarter of 1976. Data from this plant provided scale-up comparisons with smaller units on product yield, coal concentration in the feed slurry and the effects of mixed-phase (hydrogen plus coal) preheating. Spent solvent feedstock was produced for hydrotreating experiments. Operational problems that had previously limited run lengths were overcome. These included slurry sticking to the walls of the mixing vessel, plugging of the slurry preheater, plugging of the liquefaction reactor with large particles, and wear on valves and nozzles. The high viscosity of the vacuum flash tower bottoms makes pumping difficult and produces plugging.

A set of laboratory studies on hydrogenation of the recycle solvent determined the effect of mass velocity on catalyst utilization and compared the activity of several catalysts. The decrease of catalyst activity with time was studied and methods used to restore activity investigated. A one-year catalyst lifetime can be obtained by increasing the operational temperature as the catalyst ages, a technique routinely used in the petroleum industry.

Engineering Design Studies

On the basis of data obtained from the processing of Illinois #6 and Wyoming subbituminous coals in the 1-ton/day pilot plant, the issues considered to be critical for the design of the 250-ton/day pilot plant were identified as:

1. preheating and drying of the liquefaction reactor feed slurry;
2. erosion and entrainment in the vacuum flash separator;
3. corrosion resistance of metals and erosion of pressure letdown valves; and
4. design of an onstream solids withdrawal system to prevent reactor plugging by large solids particles.

Heating of the coal feed slurry produces complex heat transfer problems that are difficult to model mathematically. Consequently, scale-up of the heater design to a larger size is difficult. Scale-up of the vacuum flash separator is complicated by the entrainment of higher boiling liquids and solids in the lighter boiling liquid vapors. The tower diameter and size of the transfer line to the vacuum tower were varied to study erosion and entrainment. Erosion by the abrasive slurry is a design problem requiring features such as tungsten carbide coatings in the pressure letdown valves between the high-pressure liquefaction reactor and the atmospheric fractionation tower. A cold-flow model of the liquefaction reactor was used to identify design parameters associated with particle size distribution and withdrawal rate and frequency.

Design of the 250-ton/day Pilot Plant

The design of the 250-ton/day pilot plant was based on information obtained from the smaller 50- and 100-lb/day and 1-ton/day pilot plants, and was chosen with the subsequent requirements for scale-up to a commercial size plant in mind. The size of the pilot plant was kept to a minimum consistent with commercial scale-up practices in the petroleum industry. The engineering scale-up and operability criteria for resolving the critical technical areas determined the size of the unit. The critical technical areas were identified as: possible coking in the slurry preheater, flow distribution and stability in the liquefaction reactor, and potential coking and entrainment in the vacuum distillation step.

The planning and conceptual design of the large pilot plant began in mid-1974, and was completed in mid-1977. Contract award was made in the fall of 1977, and construction began in May 1978. The initial plan operational tests began in the summer of 1979, with the first coal processing scheduled for April 1980 as listed in Table 2-1. The test schedule is shown in Table 2-2 with operations scheduled on three coals: Illinois #6, Wyoming subbituminous and Texas lignite. Operations will be aimed at confirming scale-up criteria and determining the reliability and flexibility of the different process areas. If the operation of the large pilot plant from 1980 until 1982 is successful, it will confirm the design bases and prove the engineering scale-up information necessary to design the next step in the commercialization sequence.

Table 2-1. ECLP Test Program:
Summary of Selected Development Programs

Liquefaction Product Fractionators
   Evaluate applicability of petroleum distillation correlations
   Determine fouling, entrainment and foaming characteristics

Liquefaction Reactors
   Develop design correlations

Fluid Flow—Coal-Derived Liquids and Slurries
   Establish design correlations for saltation and pressure drop

Heat Transfer—Coal-Derived Slurries
   Determine coking rates in furnace tubes
   Develop heat transfer correlations
   Measure erosion of serpentine coal furnace tubes

Valves and Instruments in Slurry Service
   Establish reliable designs for isolation and check valves
   Determine design characteristics of pressure letdown valves
   Develop reliable level, flow and pressure (and ΔP) measurement capabilities

Pumps in Slurry Service
   Establish design criteria and correlation for both reciprocating and centrifugal pumps
   Develop reliable mechanical seal designs

Materials of Construction
   Determine corrosion/erosion rates for base and alternative materials

Air/Noise and Water Pollution Control
   Measure pollutant levels for use in future permit applications

MAJOR LIQUEFACTION PROCESSES 19

Table 2-2. Schedule for 2.5-year ECL Test Program

	Time (mo)
Operations with Illinois No. 6 Coal	15
Shakedown operations: solve operating problems, turnaround for modifications	5
Initial sustained run: demonstrate ability to operate for a prolonged period; prove out ECLP as a R&D tool	2
Exploration of operability limits: determine equipment limitations; investigate alternative conditions/configurations	4
Long-term operability: demonstrate capability to operate for an extended period and obtain time dependent data; simulate commercial operating environment	3
Turnaround	1
Operations with Subbituminous Coal—9 Months	
Shakedown and initial testing: solve operating problems; evaluate operational differences (comparison to Illinois No. 6).	3
Exploration of operability limits: same as Illinois No. 6	2
Long-term operability: same as Illinois No. 6	2.5
Confirmation of controlled scaling: chemically clean liquefaction equipment; inspect equipment (before and after cleaning)	0.5
Turnaround	1
Operations with Third Coal	6
To be developed when coal is identified. Approach will be similar to subbituminous coal program.	

A 112-ton/day (101-metric-ton/day) Flexicoking prototype unit will also be utilized to apply the process to coal liquefaction bottoms. Modification of the existing petroleum facility will begin in 1980, with completion scheduled for early 1981. Operations are planned for 18 months on bottoms from two coals generated by the 250-ton/day liquefaction pilot plant.

Integrated Research and Development Program

The EDS project is being developed with an integrated R&D program that is formulated to minimize the cost and risk in developing the process from laboratory studies to a state of commercial readiness. This approach includes an R&D program based on bench-scale studies, small pilot-unit operation, and engineering/design studies integrated with the operation of the 250-ton/day pilot plant. Operations of the pilot plant will define the process and equipment design bases for the pioneer commercial plant in the areas of liquefaction, distillation, solvent hydrogenation, bottoms processing and product quality. A close interaction between the engineering/design study team and the R&D effort will produce a smooth transition to the pioneer commercial plant with all risks and processing costs known.

An important part of the development of the donor solvent process to a state of commercial readiness is the integrated R&D program that uses bench-scale laboratory research, small pilot plants and engineering/design studies for larger plants. By using this strategy, costs are minimized and the program is able to respond to the critical development issues in a flexible fashion. The EDS program operates small pilot units at the 50-lb/day (23-kg/day), 100-lb/day (45-kg/day), and 1-ton/day (0.9-metric-ton/day) capacities, which are integrated with the 250-ton/day pilot plant, the construction of which is nearing completion. Information from the smaller units is used to predict yields from different feedstocks and processing conditions, and product quality for the 250-ton/day plant. The flexibility in plant operations is demonstrated on the smaller scale before use on the larger units. A 2-bbl/day Flexicoking pilot plant and four solvent hydrogenation units are also a part of the program.

The liquefaction pilot units are continuously operating integrated units with liquefaction reactor, solvent hydrogenation reactor, and product fractionation facilities. Extensive process development studies have been performed on the 50- and 100-lb/day units in studies of coal feed, recycle solvent quality, liquefaction reactor conditions and solvent hydrogenation conditions. The 1-ton/day unit provides scale-up data, products for further process studies, and trials of different plant processing options. Almost five-hundred 24-hr, steady-state runs have been made on the smaller units, and almost 200 runs on the 1-ton/day unit for a total of more than 30,000 hours of pilot plant operation. Continuous operation periods of two- to three-week durations have also been achieved.

The solvent hydrogenation studies have included operation of four independent fixed-bed reactor units (with 20- to 30-cm^3 catalyst volumes) and, in addition, the reactor units that are associated with the liquefaction pilot plants. Studies were conducted in the independent units on spent solvent from Illinois and Wyoming coals produced from the 1-ton/day pilot plant. Several catalysts have been evaluated, catalyst lifetime has been determined and process variables have been investigated.

By hydrogenating the donor solvent separately from the liquefaction, catalyst deactivation is considerably reduced because the catalyst is not exposed to contaminants that are in the coal. By increasing the temperature of the reactors slowly with time, the slow deactivation of the catalyst that does occur is offset and the composition of the donor hydrogen components is kept at the required level. After an initial 20 weeks of operation, a temperature increase of 0.13°F/day (0.07°C/day) will maintain satisfactory solvent hydrogenation reactor conditions with a predicted catalyst life exceeding a year.

Adaption of the Flexicoking process from petroleum residuals to coal liquefaction bottoms has included separate studies of the coking and gasification steps and the integration of both steps in a separate 2-bbl/day Flexicoking pilot plant. More than 900 hours of integrated coking/gasification operation have been logged to investigate the special characteristics of the liquefaction bottoms feedstock (e.g., ash content, solids content, pumpability and stability).

Recovery of liquid products from the vacuum bottoms has shown that Illinois coal bottoms produce about 2% more liquids than bottoms from Wyoming coal. As the coal liquefaction conditions are varied to produce larger percentages of bottoms as a fraction of the feed coal, the liquid recovered from the coker varies linearly with the quantity of the bottoms feed.

The results that have been obtained from pilot plant operation on Illinois bituminous coal and Wyoming subbituminous coal show that liquid yield increases with residence time. However, the Illinois coal has shown a greater thermal cracking of liquids to hydrocarbon gases as residence time is increased, while the Wyoming coal produces increased liquids without the gas production. Another advantage to longer processing residence time for the Wyoming coal is that the viscosity of the vacuum bottoms is decreased, which lowers the difficulty in pumping the bottoms from the vacuum tower to the Flexicoking unit. Decreases in viscosity of more than 95% have been obtained by increasing the residence time by a factor of four.

Types of Coal

Processing results have shown that the EDS process can be applied to a wide variety of coal types, including bituminous, subbituminous, and lignites. Illinois #6 bituminous coal and Wyoming subbituminous were the initially specified project coals. Other coals have undergone evaluation in small pilot plants. The 250-ton/day pilot plant is scheduled to process three coals during its operating period: Illinois #6, Wyoming subbituminous and a third coal to be selected.

An analysis of the coals that have been processed in the pilot plant is given in Table 2-3, and their liquid yields is given in Table 2-4. Longer residence times in the reactor increase conversion of the coal to liquids, but also increase hydrocracking of the liquids to gas. As a result, there is an optimum processing time in the liquefaction reactor to maximize

Table 2-3. Analyses of Feed Coals Run in RCLU

	Illinois No. 6 Bituminous		Pittsburgh Seam Bituminous		Australian Black Wandoan	Wyoming Sub-bituminous Wyodak	Texas Lignite Big Brown	North Dakota Lignite Indian Head
	Monterey No. 1	Burning Star No. 2	Ireland	Arkwright				
Elemental Analyses (dry wt %)								
Carbon	70.1	70.4	74.0	78.4	59.8	68.5	62.0	63.8
Hydrogen	5.1	4.9	5.2	5.4	5.0	4.9	4.8	4.7
Oxygen (by difference)	10.6	9.9	6.3	5.1	13.4	17.2	14.5	19.2
Nitrogen	1.2	1.2	1.2	1.5	0.7	1.1	1.1	0.9
Sulfur	4.1	3.1	4.3	2.3	0.3	0.5	1.2	1.2
Ash	8.9	10.5	9.0	7.3	20.8	7.8	16.4	10.2
Total	100.0	100.0	100.0	100.0	100.0	100.0	100.0	100.0
H/C Atomic Ratio	0.87	0.84	0.84	0.82	1.01	0.86	0.92	0.88
Ash (SO_3-Free)	8.8	10.2	8.8	7.0	20.8	6.6	14.0	7.6
Total Oxygen	15.1	14.9	10.0	8.5	26.3	23.4	23.8	26.0
Equilibrium Moisture (wt %)	14.0	10.4	2.1	1.8	10.5	29.0		33.6
Proximate Analyses (dry wt %)								
Volatile Matter	42.1	39.0	39.1	36.8	44.6	45.5	44.4	44.1
Fixed Carbon	49.0	51.2	51.9	55.9	34.6	46.7	39.2	45.7
Ash	8.9	10.5	9.0	7.3	20.8	7.8	16.4	10.2

Table 2-4. Highest Liquid Product Yields for EDS Program Coals at 1500 psig (RCLU Data)

	Coal							
	Arkwright	Ireland	Burning Star	Monterey	Wandoan	Wyodak	Big Brown	Indian Head
Liquefaction Conditions								
Temperature (°F)	840	840	880	800	840	840	840	840
Residence Time (min)	100	100	25	140	40	100	25	40
Number of Conditions Investigated	2	10	7	16	6	10	4	3
Liquefaction Yields (lb/100 lb dry coal)								
Hydrogen	−4.2	−4.6	−3.4	−4.6	−3.1	−4.8	−3.1	−4.3
Water	4.6	6.0	8.2	9.8	10.6	15.1	10.4	17.5
Carbon Oxides	1.1	1.4	1.5	0.6	3.2	5.8	6.8	7.9
Ammonia	0.7	0.6	0.6	0.7	0.3	0.5	0.4	0.6
Hydrogen Sulfide	1.8	3.2	2.4	3.4	0.2	0.5	0.7	0.4
C_1-C_3 Gas	13.5	13.5	9.5	9.0	7.1	10.1	6.2	6.8
C_4-1000°F Liquid	29.9	32.7	30.4	36.1	27.7	30.9	28.0	28.1
1000°F Conversion (lb/100 lb dry coal)	47.4	52.8	49.2	55.0	46.0	58.1	49.4	57.0
Liquid Product Selectivity (wt %)								
C_4-400°F Naphtha	73.2	71.6	48.0	62.7	66.7	80.1	57.6	68.5
C_4-1000°F Liquids								

liquid yield, although additional liquids may be obtained by recycling or separately processing the bottoms by Flexicoking. The types of coal that produce high yields have low ash content, and contain high quantities of volatile matter, sulfur and reactive fractions.

The percentage of volatile material in the coal is the most important coal property used to predict the liquid yield. In performing a single variable regression analysis, 88.5% of the variation about the mean can be removed using this single parameter.

Bituminous coals give a total liquid yield of 43–45%, subbituminous coals give 40% yield, and lignites produce 33–35% liquid. The products that are produced have higher levels of nitrogen than similar fractions of petroleum and have sulfur levels in proportion to the original sulfur content of the coal. Both sulfur and nitrogen concentrations can be reduced by subsequent processing using standard petroleum hydrotreatment processes. The pilot plant operations have shown that the younger subbituminous coals and lignites are more difficult to process due to the higher oxygen and organically associated calcium content. Formations of calcium carbonate deposits on the walls of the reactor have proven to be an operational problem. If not removed or prevented, these deposits can plug the reactor and foul downstream equipment. In the 250-ton/day pilot plant, an attempt will be made to control this problem with strainers upstream of critical equipment, such as valves, instruments and pipe bends. Large, solid particles will periodically be withdrawn from the liquefaction reactor, and chemical cleaning will be used during turnaround time. This mechanical method is preferred due to simplicity and more favorable economics. However, if it does not work satisfactorily, pretreatment of the crushed coal with sulfur dioxide converts the calcium into calcium sulfate, which does not form reactor deposits under the EDS operating conditions. Another problem in processing the younger coals is the high viscosity of the coal liquefaction bottoms. The viscosity is a direct measure of the difficulty in pumping the bottoms from a vacuum fractionator into a coking or gasification reactor. The viscosity of the younger coals can adequately be reduced to pumpable levels by using longer liquefaction residence times.

Liquefaction Product Market Studies

The EDS process produces a spectrum of liquid products from butanes through 1000°F (538°C) and higher coker liquids. Operation of the process to produce the most profitable output is a problem that is complicated by the fact that coal-derived liquids, as produced by product fractionation, are frequently not directly substitutable for analogous petroleum products. Furthermore, predicting markets and product prices a decade or more in advance is difficult when decisions must be made at present on the processes to be used for the next generation of plant. Product market studies must anticipate future energy consumption patterns and the end use equipment assumed to operate in the commercial plant time frame.

The principal liquid products to be made in the EDS commercial plant are liquefied petroleum gas, naphtha and middle distillate to heavy fuel oils. The petroleum gases are a finished product; the naphtha is an unfinished feedstock for a downstream refinery/petrochemical plant; and the fuel oil is suitable for sale to customers having the facilities to receive, handle and burn it. For Illinois coal, operation of the process at the preferred conditions would produce 37% of the total liquids in the form of 350°F (177°C) naphtha. A distillate fuel oil boiling in the 350–650°F (177–343°C) range would represent 24% of the liquid output, while 39% would be 650°F and above, heavy fuel oil. The distillate fuel oil and heavy fuel oil have been used in combustion tests to determine if environmental requirements are met and combustion properties are acceptable. Additional treatment, using available technology, will be required to remove nitrogen from the fuel oil to limit oxides of nitrogen emissions.

The use of EDS coal naphtha for reforming into high-octane gasoline blending feedstock will require single- or two-stage hydrotreating facilities to eliminate reforming catalyst poisoning by sulfur, oxygen and nitrogen in the naphtha. However, these facilities are available at a price range which is competitive, and the gasoline market should be strong. The extent to which the heavy fuel oil market will receive competition from the direct combustion of coal is difficult to predict. Testing of the fuel oils is being performed to determine combustion and air quality emission characteristics to determine potential utilization as diesel and jet fuel, or home heating oil and industrial fuel oil. High concentrations of aromatic compounds can cause fuel stability and combustion problems, requiring equipment modifications in some existing equipment and applications.

Since all of the products from the 250-ton/day pilot plant will be utilized in the nearby Exxon refinery at Baytown, TX, laboratory studies have been undertaken to test compatibility with the petroleum products. Some blends are not miscible, while others can only be mixed in small quantities to avoid producing a product with unacceptable specifications. Special handling or upgrading steps may be required to process commercial plant output if utilized in pertoleum processing facilities.

26   LIQUEFACTION

Because half of the coal feed emerges from the reactor as vacuum bottoms, the disposition of this high-ash product is an important process consideration. This stream, which contains one-third to one-half of the available carbon in the feed coal, will be used to achieve plant hydrogen and fuel balances for the overall process, to convert the available carbon into a useful form, and to minimize costs. Exxon Flexicoking is a commercial petroleum process that employs an integrated coking/gasification sequence in circulating fluidized beds. The heavy vacuum bottoms are fed to the Flexicoking unit with air and steam to produce additional distilled liquid products and a low-Btu fuel gas for the process furnaces. All of the organic material is recovered as a liquid product or combustible gases. Other possibilities that are being investigated include the production of hydrogen and fuel gas, using either the Texaco or Shell/Koppers partial oxidation processes. Another possibility is recycling the bottoms through the donor solvent process to produce additional liquids.

The technical issues on the disposition of the bottoms requiring further investigation are:

Coal Liquefaction Bottoms Properties	Process Development Issues
High ash/solids content	Gasifier slagging; particulate generation/control
High viscosity	Bottoms pumpability
Thermal instability	Feed control and distribution

For Flexicoking, a principal issue is the impact of the high mineral matter content on particulate generation/control and gasifier slagging. Resolution of these issues has led to expansion of the program to operate a 112-ton/day (101-metric-ton/day) prototype Flexicoking unit. Construction will be completed in early 1981, and operations begun on bottoms from two coals generated by the 250 ton/day liquefaction pilot plant.

Modeling of the EDS Process

A mathematical modeling effort has been a part of the hydrogen donor solvent development program. Simple empirical models have been developed to predict the product outputs of the developments units as a function of process conditions as well as a more fundamental model that analyzes the chemical reactions involved in the donor solvent process. The fundamental model permits the developers

MAJOR LIQUEFACTION PROCESSES 27

1. to correlate and predict differences in liquefaction yields due to variations in feed coals;
2. to gain a better understanding of the chemistry of coal liquefaction; and
3. to predict more precisely the product distribution from coal liquefaction.

A complete model that would define the reaction paths of the individual molecular species and calculate the reactions over a wide range of process conditions and hydrogenation feedstocks would be very complicated. By lumping similar reactions into classes of compounds, rate constants and activation energies have been determined that can predict the concentrations of the key components for typical reactor and catalytic hydrogenation conditions. The chemical species and reactions chosen are based on the known chemistry of coal, coal liquids and donor solvent. From these reactions, rate equations are derived and integrated numerically to give concentrations of chemical species as a function of reaction time. The results of the calculations are compared to pilot-plant data, and the rate constants, activation energies and equations adjusted to give a best fit of the data.

The program calculates a material balance summary for nine processing steps and prints out the contents of each stream. The streams are characterized in terms of the nine component fractions shown with representative percentage weight yields.

Hydrogen	(−3.7)
C_1–C_3 gas	5.3
Chemical gasses (H_2S, NH_3, CO, CO_2)	4.7
C_4–400°F (204°C) naphtha	17.9
400–700°F (204–371°C) liquids	7.6
700–1000°F (371–538°C) liquids	7.6
1000°F+ (538°C+) vacuum bottoms	52.0
Dry feed coal	(−100.0)
Water	8.6
Total	0.0

The yields are computed as a function of the following seven independent variables:

1. liquefaction temperature,
2. liquefaction residence time,
3. solvent temperature,

28 LIQUEFACTION

4. solvent hydrogen partial pressure,
5. solvent hydrotreater space velocity,
6. solvent/coal ratio, and
7. composition and amount of 1000°F- (538°C-) liquid in bottoms.

Exercising the model has duplicated process development unit results and verified one of the strong features of the EDS process: the maximum yield of 400–1000°F (204–538°C) boiling temperature liquids is insensitive to process temperature. At any temperature in the range 800–860°F (426–460°C) a liquid yield of approximately 14% can be obtained if the proper liquefaction residence time is chosen. Another positive feature that can be demonstrated is that a wide range of product slates may be obtained by optimizing conditions in the liquefaction and solvent hydrogenation steps. This wide range of products is obtainable without employing extremes of operating conditions and is readily attainable using conventional equipment used in the petroleum processing industry.

EDS Process Alternatives Linear Programming Model

A linear programming model based on pilot-plant experience has been used to study process alternatives and their complex interactions to maximize income from sale of products, while minimizing processing costs and maintaining an optimum blend of product production. The basic relations between product yield and reactor treatment conditions are programmed to evaluate alternative conversion processes for generating the hydrogen and fuel gas required for liquefaction. The possibilities studied included air and oxygen Flexicoking, partial oxidation and moving bed gasification of liquefaction bottoms or additional raw coal. The costs and yields of these processes, and restrictions due to material balance, feed availability, product demands, equipment cost and capacity are used by the model to optimize process options. The features of the model include:

- computerized weight and energy balances;
- simultaneously balanced hydrogen, fuel gas and bottoms disposition;
- definition of the utilities required for each case, with steam and electric power generation optimized;
- investment estimate calculated for each case;
- each case balanced to minimum overall cost;
- parametric examination of key variables, such as relative product values and volumes;

- feedstock flexibility, with coal available to generate hydrogen, fuel gas, power and steam as well as feed to liquefaction;
- selection of optimum product slate as a function of price structure (potential products include naphtha, liquefied petroleum gas, low-sulfur fuel oil (several grades), high-, low- and medium-BTU fuel gas and vacuum bottoms);
- modeling of by-product processes for sour water stripping and recovery of ammonia, phenols and sulfur; and
- utility costs, including electric power, raw water, recirculating cooling water, demineralized boiler feed water, 600- and 150-psig steam generation costs and steam turbine drivers substituted for electric motors.

The process alternatives model has been used to defend the optimum plant configurations, to select the equipment sizes and to determine capital and operating costs of the configurations.

Conceptual Commercial Plant Designs

In 1975–1976 the Illinois #6 Coal Base Case design was established. In 1978–1979 the design was updated and revised for a plant processing 30,000 ton/day (29,528 metric ton/day) of coal into 65,000 bbl/day of liquefied petroleum gas, naphtha and fuel oil. A sensitivity analysis was also conducted, which produced hydrogen from the partial oxidation of vacuum distillation tower bottoms, instead of steam reforming of C_2–C_3 light gases as had been used in the base case design and 1978–1979 revision. Elimination of the high energy consumption of steam reforming increased the plant thermal efficiency from 55.6% for the base case to 63.6% for the revised design. The C_2–C_3 gas was available for sale as a high-Btu pipeline product, and plant steam generation requirements were significantly reduced. The plant configuration that results is shown in Figure 2-2, consisting of four parallel liquefaction reactors serviced by two donor solvent hydrogenation units. Two Flexicokers, three oxygen plants and six partial oxidation gasifiers service the facility. A new base case design using Wyoming subbituminous coal is now under study. Limited screening studies have been performed on plants using Pittsburgh #8 bituminous coal and Texas lignite from the Big Brown mine. A definitive preliminary design study has not been completed. The current data are subject to revision.

Because the data are not available for the Illinois sensitivity case, the economic analysis for the base case and design update will be presented. Tables 2-5 and 2-6 present the total cost to erect the plant by

30 LIQUEFACTION

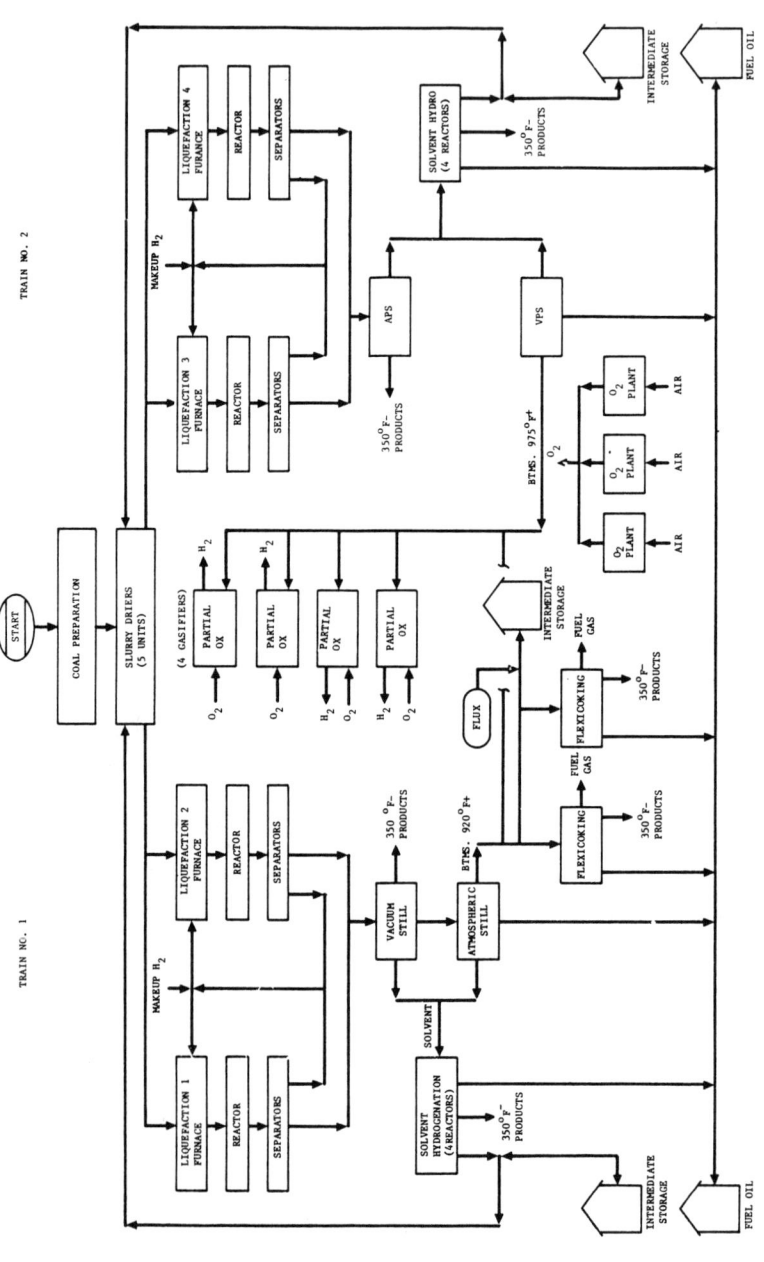

Figure 2-2. EDS commercial plant study design update. Illinois coal market flexibility sensitivity case, configuration of units (source: EDS Commercial Plant Study Design Update, 1979).

major sections while Table 2-7 shows the annual operating expenses. The major reasons for the doubling in cost are:

1. changes in scope and capacity—51%
2. design developments—27%
3. price escalation and three-year startup date delay—13%
4. design/estimating developments—16%

The assumptions for the economic analysis are given in Table 2-8. Table 2-9 gives the daily material and product balance.

Pilot-Plant Operations Description of the 1-ton/day Pilot Plant

The coal that is fed to the liquefaction plant is handled and prepared in an inert nitrogen atmosphere. The dried coal contains less than 1%

Table 2-5. Exxon Donor Solvent Commercial Plan Study Design Update: Construction Cost[a] Estimate (10^6 \$, 1978 third quarter)

Construction Cost	
Onsite Units	\$1018
Offsite Units	\$ 521
Total Construction Cost	\$1539
Indirect Costs	\$2234
Total Construction and Engineering Cost	\$3773
Project Life	30 yr
Capacity Factor	85%
Thermal Efficiency	55.6%

[a]Excluded costs: inflation, project and process contingency, coal mine development, land (1450 ac), product pipelines and right-of-way, planning and feasibility studies, and external power generation plant.

Table 2-6. Investment Expenditure Schedule

Year[a]	% Equipment Cost	% Startup Cost	% Yearly Production
−4	4.5		
−3	9.5		
−2	18.5	0	
−1	25.1	5	
0	23.5	15	0
+1	14.5	45	25
+2	4.5	35	81
+3	0	0	100

[a]First train completed in year zero; second train completed in year +1.

32   LIQUEFACTION

Table 2-7. Exxon Donor Solvent Commercial Plant Study Design Update: Annual Operating Expenses (10^6 $ 1985)

Coal	316
By-products	
Sulfur	(22)
Ammonia	(18)
Utilities	
Electric Power at $.03/kWh	85
Water	NA
Catalyst/Chemicals/Waste Disposal	23
Salary and Related Costs	82
Other Operating Costs	155
Subtotal	621
Capital Cost Recovery @ 15% Discounted Cost Factor Return	326
Total Annual Cost	947

Table 2-8. Exxon Donor Solvent Commercial Plant Study Design Update: Assumptions for Computing Annual Operating Expenses (10^6 $/yr)

Operating and Maintenance Personnel	
Hourly Wage Workers	1071
Professional Salaried	269
Contract Maintenance	317
Total	1657
Salaries and Related Costs	
Wage Earner	20,000
Salaried	27,000
Benefits	32.5% of salaries and wages
Overhead	20% of salaries and wages
Coal Cost (cleaned coal at mine)	$22/ton ($24.2/metric ton)
Taxes	50% of gross profit

moisture and is crushed to 30–100 mesh. The slurry mixture of coal and recycle donor solvent is continuously prepared at atmospheric pressure and temperatures less than 125°F (52°C). The design of the slurry preheater was one of the most difficult in the process, and one which is difficult to scale-up as plant size increases. Separate phase heating of the hydrogen gas and slurry is less desirable than mixed phase preheating because preheater coking tendencies are minimized in the mixed phase mode. Preheater design is complicated by inadequate knowledge of the viscosity characteristics of the slurry, film temperatures, residence time, partial vaporization and variations in feed slurry composition.

The donor liquefaction reactor system uses an upward plug flow design with operating conditions of 800–900°F (427–482°C) and 2000 psig (141 kg/cm^2) pressure. The coal feed reacts with gaseous hydrogen

MAJOR LIQUEFACTION PROCESSES

supplemented with hydrogen from the donor solvent, and is broken down into gases, light hydrocarbon liquids, heavy oils, unreacted coal and ash.

Liquefaction product depressurization is performed in two stages. The high-pressure separation system uses a lock valve to depressurize the slurry to the product separator pressure of 1500 psig (105 kg/cm^2). The high-pressure product gases, light hydrocarbons and water are cooled before separation of the hydrocarbons and water. The cooled high-pressure liquefaction gases are subsequently processed in the recycle gas cleanup system. The low-pressure separator operates at 100 psig (7 kg/cm^2) and 700°F (371°C). Flash gases and light hydrocarbons are separated from the heavier hydrocarbons, which are recovered for feed to the hydrotreater.

From the low-pressure separator, the product mixture is fed to the vacuum flash system which operates at 25 mm Hg pressure and up to 750°F (455°C). The mineral matter and 1000°F+ (538°C+) liquefaction bottoms product are separated from the lower-boiling fractions. The bottoms level is monitored with a nuclear level detector, and a gear pump is used to remove the high viscosity bottoms from the vessel where it is cooled and solidified. The viscosity of the vacuum flash bottoms is also continuously monitored. A side stream of 700–1000°F (371–538°C) material is removed from the lighter material, cooled and sep-

Table 2-9. Exxon Donor Solvent Commercial Plant Study Design Update: Material and Product Balance

Input	
Coal (as received, Illinois #6 Monterey #1 mine)	
to liquefaction units	30,000 tons (27,000 metric tons)
to boilers	NA
Electric Power (assuming all electric drives)	184 MW(e)
Raw Water	NA
Products	
Propane LPG	2,610 bbl
Butane LPG	2,460 bbl
Naphtha	23,200 bbl
Fuel Oil	36,970 bbl
Sulfur	996 tons (903 metric tons)
Ammonia	132 tons (120 metric tons)
Phenol	480 bbl
Stream Flows	
Spent Solvent (per train)	98,600 bbl
Vacuum Bottoms (per plant)	53,000 bbl
Phenolic Water (per train)	42,900 bbl

arated from noncondensable gases in a packed column. The condensed and cooled overhead flash products are accumulated and pumped to a hydrotreater for subsequent upgrading. The hydrotreater reactor has a catalyst bed with several quench hydrogen and product oil injection points spaced along the length for temperature control. Water is removed from the hydrotreater product before fractionation.

High-pressure product gases from liquefaction and hydrotreating are scrubbed with a water solution of monoethanolamine in the recycle gas cleanup system to remove acid gases. The gas scrubber operates at 1000–1500 psig (70–105 kg/cm^2) at 150°F (66°C). The scrubbing solution concentration, solution feed rate and scrubber dimensions are chosen such that greater than 99% of the entering hydrogen sulfide and carbon dioxide gases are removed. The scrubbing solution is recycled through a regenerator.

The recycle gas compression system is designed to handle the combined streams from recycle scrubbing plus fresh hydrogen makeup of 95% minimum hydrogen purity. The low-pressure gas scrubbing system recovers light condensable hydrocarbons from low-pressure gas streams before discharge. Rich oil from the scrubber is pumped to the naphtha/solvent fractionator. The naphtha/solvent fractionator removes the light front-end components from the hydrogenated solvent stream and separates water from the naphtha.

The spent donor solvent emerges from fractionation deficient in hydrogen. It is catalytically hydrogenated to replenish the donor hydrogen in a process step that is separate from the coal processing operations. The replenished solvent is then recycled and mixed with raw feed coal before entering the liquefaction reactors. The catalytic hydrogenation step is accomplished using conventional petroleum hydrotreating technology and utilizes commercially available petroleum hydrogenation catalysts.

Operation of the 1-ton/day Pilot Plant

Numerous problems were encountered during the operation of the 1-ton/day pilot plant. During its initial operation in the first quarter of 1976, yield data were obtained that permitted direct comparison with the yields obtained from the smaller integrated pilot plants. Several runs investigated the coal concentration in the feed slurry and the effects of mixed phase (recycle hydrogen plus coal slurry) preheating. Spent solvent was generated for bench scale hydrotreating experiments.

MAJOR LIQUEFACTION PROCESSES 35

Following a scheduled six-month shutdown, the plant was restarted in the fall of 1976 to establish operations with good material balance. A run with a subbituminous (Wyodak) coal lasted 108 hours without experiencing plugging problems from calcium carbonate formation, and explored the operability problems associated with these coals.

The yield studies showed good agreement with data from the smaller plants. The degree of conversion was the same, and showed identical response with changes in temperature and liquefaction reactor residence time. No significant differences in product yields could be attributed to plant size. The coal concentration was shown to have a significant economic effect on the liquefaction system design. As the coal concentration in the slurry increased, the quantity of recycle solvent to be processed decreased. Thus, the solvent recovery and hydrogenation systems are smaller, and a lower investment cost in plant equipment is required.

If the solvent-to-coal ratio declines below a critical limit, the liquefaction yields decrease significantly. The coal concentration is also limited by slurry pumping and coking in the preheater. Maintaining the highest coal concentration in the slurry consistent with adequate yield reduces the size of pumps and pressure vessels.

Mixed-phase preheating of the feed slurry and hydrogen is preferred over separate heating of the hydrogen and feed. Due to the problems of scaling up the thermal analysis of the three-phase flow occurring in a larger preheater, the final acceptability of mixed-phase preheating must be determined with data taken when the 250-ton/day pilot plant begins operation.

Some of the problems encountered in the pilot plant operation and mechanical improvements made to solve them are as follows:

- Coal would accumulate on the walls of the slurry mixing vessel and on top of the slurry. Centrifugal mixing pumps were added to agitate the slurry.
- Plugging of the slurry preheater tubes occurred due to coking. Tube diameters were reduced, and better control was exercised over the unit operation.
- Since the upflow tubular liquefaction reactor did not have a high enough velocity to transport the largest particles, plugging occurred. A solids withdrawal system was installed to permit periodic removal.
- Liquefaction reactor products are withdrawn through a high-pressure–high-temperature liquefaction separator. The slurry is depressurized from 1500 to 100 psig (105 to 7 kg/cm^2) through control valves. The abrasive nature of the slurry and large pressure

- drop produces wear on the valve stems. Hard coatings have been used to extend valve lifetime.
- The next step is a low-pressure–high-temperature liquefaction separator that is also subject to wear from entrained solids. Hard coatings have been applied to surfaces subject to abrasion. Solids have been responsible for valve plugging that has terminated runs.
- Because of the small differences in specific gravity between the oils (0.96) and sour water (1.02), the condensate in the liquefaction oil-water separator is difficult to separate into oil and water phases. An interface detector that monitors the interface between oil and water was added to assist the separation.
- The primary vacuum flash tower separates the lower-temperature hydrocarbons from the heavy vacuum bottoms. The pumps used to remove the heavy vacuum bottoms suffered from wear and had short lifetimes. These problems have been eased by changing to new pump designs and closely controlling the viscosity of the bottoms produced by the liquefaction process. A circulating hot oil heating system was installed on the bottoms pumparound loop to provide uniform heating and eliminate plugging due to coking.
- A hot oil heating system also eliminated plugging problems on the secondary vacuum flash tower bottoms pumparound loop.
- Originally the temperature in the hydroheaters was regulated by injecting hydrogen gas into hot areas. Substitution of liquid product from the reactors was used to replace recycle hydrogen gas as the quenching medium. Operation was smoother; pressure surges decreased and the level of temperature control improved.

Additional equipment was installed for special engineering studies:

- A special sectionalized slurry preheater was installed to study viscosity changes and coking that occur during the heating of coal slurries. Pressure taps and sample points between the five sections enable data to be obtained on pressure drop and physical properties of the slurry.
- A high-temperature–high-pressure slurry viscometer was used to measure the viscosity of coal feed slurries and vacuum tower bottoms as a function of temperature, pressure and shear rate. The rheological data are used to design piping, heat exchangers, furnaces and pumps.
- Piping test sections and corrosion test coupons were used to measure the corrosiveness and erosiveness of the slurry, liquid hydrocarbon products and sour-water streams.
- Operations with Wyodak coal produced high-viscosity bottoms resulting from higher reactor temperatures because of the different reactivity of the coal.

Project Assessment

Extensive bench scale and small pilot plant testing has been performed on the Exxon Donor Solvent process since 1966. The 250 ton/day integrated process operation pilot plant began operation in the first half of 1980. Pilot plant operation will demonstrate process operability and establish process economics. A scale-up in size by a factor of 100 will be required to develop a commercial plant. The chief process area requiring definition is the optimum processing of vacuum bottoms. A Flexicoking pilot plant will operate in the same time period as the liquefaction plant to establish the feasibility of this process. The preliminary design and specifications of a commercial plant will be performed along with pilot-plant operations. Scale-up data and other necessary information for commercial plant design will be obtained from pilot plant operations. A decision on commercial plant development will be made during the pilot plant operation based on satisfactory technical operation and favorable economic position relative to petroleum fuels. Government policy for the support of synthetic fuel technologies is now being formulated.

H-COAL PROCESS

Process Description

The H-Coal process is a direct catalytic hydroliquefaction processing method for converting high-sulfur coal into either a heavy fuel oil that will meet environmental sulfur emission regulations or into a variety of liquid products considered equal to a synthetic crude oil. The process is characterized by an ebullating bed reactor in which a slurry of coal and recycled oil is brought into direct contact with the catalyst. The reactor uses an upward flow of liquid and hydrogen to expand the catalyst bed and distribute the slurry, gas and catalyst evenly across the reactor. The ebullating bed reactor permits a moderate accumulation of solids to be present in the system without plugging the reactors, promotes optimum catalytic activity at a uniform temperature and achieves a simple and economic equipment configuration for promoting the reactions which liquefy coal.

The H-Coal process was developed from previous work done by the developer, Hydrocarbon Research, Inc., on the liquefaction of petroleum residual oils and tar sands. The process is called H-Oil, utilizes a similar

38 LIQUEFACTION

reactor, and has been in operation on a commercial scale for many years. The H-Coal process technically is a descendant of the Bergius process for direct catalytic hydrogenation, but employs a more effective catalyst and operates at lower pressure and with reduced hydrogen consumption.

The type of output liquid produced is determined by the residence time of the coal in the reactor and the amount of hydrogen consumed. More prolonged treatment conditions convert the product into lighter-weight hydrocarbons and remove greater quantities of sulfur and nitrogen. The processing capacity of the reactor decreases as the treatment time of the coal increases. The Catlettsburg, KY, pilot plant now completing construction will be the largest coal synthetic liquids plant in the United States. It will have a capacity of 600 ton/day (545 metric ton/day) of coal when operating in the boiler fuel mode, or 250 ton/day (227 metric ton/day) when operating in the synthetic crude mode.

The H-Coal process development began in 1964 with government funds. When government funding was discontinued in 1967, the process development was continued with support from private corporations. Government sponsorship was revived in 1973, leading to construction of the Catlettsburg pilot plant. Many years of operation at the bench-scale and small process development unit level have demonstrated the H-Coal process capability to handle 14 types of bituminous and sub-bituminous coal and lignite from all geographical regions of the country.

The H-Coal process is illustrated in the flow diagram of Figure 2-3. Coal is crushed, dried and slurried with a recycle oil. The slurry is heated to a temperature below the 750°F (404°C) operational temperature of the reactor in a fuel gas–fired preheater and mixed with hydrogen. The reactor temperature is hotter because of heat released in the liquefaction process. The ebullating bed reactor brings the coal into contact with the cobalt/molybdenum catalyst, where hydrogeneration and cracking of the coal molecular structure occurs. Sulfur, nitrogen and oxygen are converted into hydrogen sulfide, ammonia and water. The hydrocarbon products produced range from heavy fuel oil to naphtha and light gases, depending on the length of time spent in the reactor. Since the hydrogenation reactions are exothermic, the reactor temperature control is maintained at the optimum level by balancing the heat input from the entering preheater slurry and the output to the product separator. Circulation of the reacting mixture of gases, liquids and solids in the reactor promotes constant reactor temperature, keeps the material in good contact with the cataylst bed and prevents plugging by solid ash material and undissolved coal.

MAJOR LIQUEFACTION PROCESSES 39

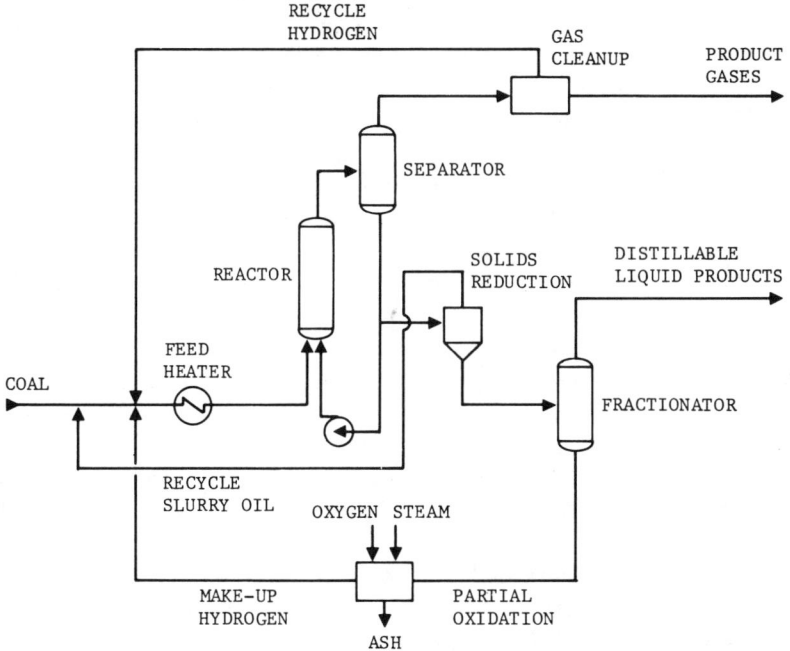

Figure 2-3. H-Coal process (source: H-Coal Technical Proposal, 1978).

The product mixture is depressurized into a product separator. A mixture of gases and hydrocarbon vapors are taken off at the top; a slurry of liquids, undissolved ash and unconverted coal is removed from the bottom. The gases are sent to a recovery plant where hydrogen is separated for recycle to the preheater. Light hydrocarbon gases are removed into several product streams, and a fuel gas is produced for use in plant process heaters and for steam generation. Ammonia and hydrogen sulfide are sent to recovery plants to produce marketable sulfur and anhydrous ammonia. The slurry is depressured to flash into a vapor fraction and a liquid containing recycle oil, unreacted coal and ash. The liquid is centrifugally separated into a high solids content and a low solids content stream in hydroclones. The low solids stream is used as recycle oil and sent to the slurry preparation system. The high solids content stream flows to an atmospheric distillation tower. The tower overhead is sent to a products fractionator with other liquid hydrocarbon streams. The bottoms from the atmospheric tower is sent to a vacuum distillation tower. The distillation tower separates hydrocarbon products from the bottoms, which consist of heavy hydrocarbon liquids,

40 LIQUEFACTION

unreacted coal and ash. The bottoms are fed into a partial oxidation unit with steam and oxygen, to produce makeup hydrogen from the process and a fuel gas. The coal ash is then rejected without having to be separated by filtration.

All process hydrocarbon liquids are separated into light gases, naphtha, distillate oil and heavy fuel oil and subsequently processed with conventional petroleum refining technology to produce a marketable product. Additional hydrotreatment steps can reduce concentrations of sulfur, nitrogen and oxygen. The product yield can be altered by varying the treatment in the reactor. More than 90% of the feed coal can be converted to liquids.

The advantages of the H-Coal technology are that it is based on a commercially proven petroleum technology that has been thoroughly tested at a size which is within the range of a commercial coal liquids plant. Much of the equipment development technology can be transferred to H-Coal, along with operating and control procedures that have been thoroughly tested. The H-Coal process has been tested successfully on many commercially important American coals. Data has been obtained in 60,000 hours of bench-scale operation and 10,000 hours of operation on a 3-ton/day process development unit.

The pilot plant now completing construction at Catlettsburg, KY, is the largest coal liquefaction plant built in the United States. Operation of the 200- to 600-ton/day facility will provide the data necessary to make a scale-up from a pilot plant to a full commercial facility using experience obtained in petroleum processing.

A proposal for a 50,000-bbl/day commercial H-Coal plant has been submitted to the DOE by Ashland Synthetic Fuels, a subsidiary of Ashland Oil. The objective of this program will be to design, plan, construct and operate a commercial coal liquefaction plant under a commercialization group that will include corporations, representatives of utilities and foreign organizations. The value of foreign oil replaced by synthetic products would exceed $700,000/day. If imported oil prices increase more rapidly than coal prices, the project could become profitable without government support, but will probably require a government subsidy.

Technical Evaluation

Historical Development and Status

The H-Coal process is a catalytic hydroliquefaction process that converts high-sulfur coal to a boiler fuel that will meet sulfur emission

MAJOR LIQUEFACTION PROCESSES   41

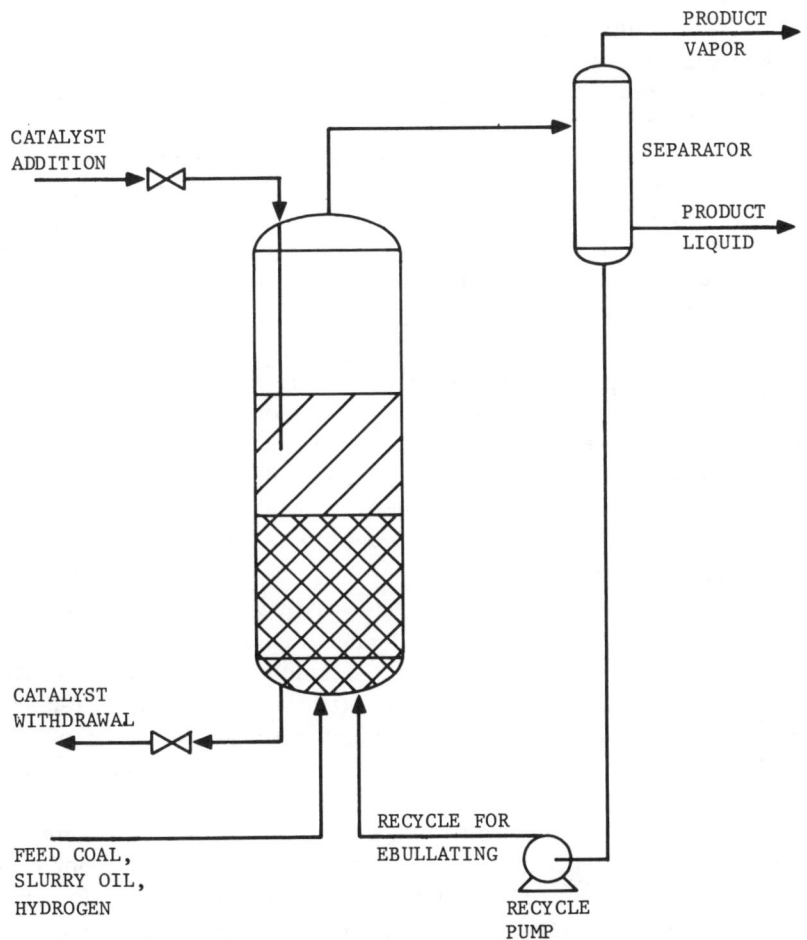

Figure 2-4. H-Coal ebullated-bed reactor (source: H-Coal Technical Proposal, 1978).

regulations or to a refinery syncrude. The first H-Coal experiments were carried out by Hydrocarbon Research, Inc. (HRI), with its own funds in 1964. The ebullated bed (H-Oil reactor) invented by HRI in the late 1950s was used to hydrogenate a coal-oil slurry.

The ebullated-bed reactor, shown in Figure 2-4, uses an upward flow of liquid (or a slurry of finely ground coal in oil) and hydrogen to expand the catalyst bed; to distribute slurry, gas and catalyst evenly across the reactor; and to suspend the catalyst particles in random motion. This ensures optimum catalyst activity by permitting the system to operate under essentially isothermal conditions and by allowing for the continuous addition and withdrawal of catalyst during operation.

The expansion of the catalyst bed permits a moderate amount of inert solids to be present in the feed, and the isothermal bed eliminates the need for quench points. Ebullating action within the bed prevents dead spots and channels from forming and makes vapor-liquid redistribution devices unnecessary.

The ebullated bed was originally developed for the catalytic hydrogenation of heavy oil extracted from Alberta tar sands. The process of hydrogenation of heavy oils in the ebullated bed to produce lighter, more useful oils is called the H-Oil process; the process of hydrogenation of coal to produce liquid fuels is called the H-Coal process.

The earliest application of the ebullated-bed reactor was the upgrading of heavy petroleum fractions. This process, developed in cooperation with Cities Services Research and Development Company, desulfurizes and demetallizes heavy fuel oil and residual oil and converts residuum oil to distillates. A commercial H-Oil unit was operated in Lake Charles, LA. Commercial H-Oil units are currently in operation in Shuaiba, Kuwait and Salamanca, Mexico.

The initial H-Coal experiments were carried out by HRI in 1964. The results obtained were sufficiently encouraging so that the Office of Coal Research (OCR) granted HRI a $2 million, 3-year development contract in February 1965. All work for this contract was in the syncrude mode (maximum distillate production) and was done on two scales. The initial experiments were carried out in 25-lb/day (11.4-kg/day) coal feed rate "bench units." These units have the advantage of low inventory, quick startup, rapid achievement of equilibrium following change of conditions and reasonably low operating labor cost. The flows, however, were too small to sustain the continuous operation of product separation facilities that produce the slurry-oil recycle streams envisioned for a commercial-scale unit. In addition, it is very difficult to control bed expansion on this scale, and an internal screen is required in the reactor. To overcome these limitations, a process development unit (PDU), capable of feeding up to 3 tons (2.7 metric tons) of coal per day, was also employed.

The bench-unit OCR program operated for slightly over 8000 hours in five different tests. These tests were as follows:

1. Illinois No. 6 Coal (6000 hours). Four experiments were involved: (1) Feed-slurry oils not derived from coal were tested for 2800 hours. Catalytic cracker decant oil and anthracene oil were used. The most significant result was an improvement in conversion when

hydrogenated anthracene oil was employed, probably as a result of a donor solvent effect. (2) Process variable studies were conducted from 1200 hours. Temperatures of 830–860°F (444–462°C), pressures of 1500–3000 psi (105–211 kg/cm^2), space velocities of 15–45 lb/hr/ft^3 (0.19–0.58 kg/hr/m^3) and other process conditions were evaluated. (3) The varying residuum content of feed-slurry oil was tested for 1000 hours. A vacuum column was used to vary the ratio of distillate to residuum in the slurry to demonstrate the effect of this ratio on yields. (4) An alternative catalyst was studied for 1000 hours. Nickel molybdenum catalyst available at that time was found to be no better than the standard cobalt molybdenum catalyst.
2. Wyodak, Wyoming Subbituminous Coal (1550 hours). Yields were determined and compared with Illinois No. 6 coal.
3. North Dakota Lignite (550 hours). A yield comparison run was made.
4. Utah D Coal. A short yield comparison run was made.
5. Two-Stage Processing. A short run was made. It employed PDU vacuum tower bottoms in place of coal feed to determine if a two-stage operation significantly enhances coal and residuum conversion.

The PDU had several short runs. Mechanical problems due primarily to the erosive nature of the coal-oil slurry were resolved. A 473-hour continuous run on Illinois #6 coal was then made, and the results provided a basis for relating bench-unit and PDU yields.

From October 1967 to March 1968, HRI continued development work with its own funds. In March 1968 the Atlantic Richfield Company contracted with HRI for an enlarged research and development program. This program resulted in significant progress in H-Coal process development and was concluded in 1970. Additional process variable studies were run on several coals in the bench units at constant temperatures and pressure in the syncrude mode. Space velocity and slurry composition were varied to determine conversion and yields for the following coals:

- Pittsburgh seam—1400 hours of operation
- Texas lignite—1200 hours of operation
- Black Mesa—1200 hours of operation
- Colorado—1400 hours of operation

In 1970 a 400-hour run was made on Australian Gelliondale brown coal. This proved to be a very reactive coal; conversions greater than 80% were achieved, with less than 20% of the yield being residuum

44 LIQUEFACTION

boiling above 975°F (524°C). In early 1971 the boiler fuel mode (residual fuel production) of operation was demonstrated in a 716-hour bench unit run on Illinois #6 coal. The results were used to predict the conditions needed to arrive at a product with acceptable residuum and sulfur contents. An additional 2000 hours on Illinois #6 coal confirmed these projections and tested other process improvements, such as the use of hydroclones to produce a high-residuum/low-solids recycle slurry oil and the effect of using a more coarsely ground coal feed.

In 1972 and 1973 HRI contracted with six major oil companies for financial support to accelerate the development of the H-Coal process to a point where commercial-scale plants could be designed and built with confidence. Under industrial sponsorship, bench-unit operations lasting almost 3000 hours were carried out with Wyodak coal to determine yields and to test such additional process features as the use of coal dried with flue gas and the behavior of regenerated catalyst. During this period, the PDU was also operated for almost 2400 hours in the syncrude mode. Recycle slurry oil was provided by continuous product fractionation and the batch use of hydroclones on the reactor-product slurry. In summary, the bench-scale unit was operated for about 1600 days. Four bituminous coals, four subbituminous coals, and three lignites were tested, with Illinois #6 and Wyodak coals being the principal coals used.

Substantial funds were spent in the development of the H-Coal process prior to support by the Energy Research and Development Administration (ERDA) (now DOE) in 1974. A summary of this funding is as follows:

- From 1958 to 1965, approximately $10 million was spent in: (1) demonstrating the technical feasibility of the ebullated reactor concept and (2) conducting initial exploratory experiments on the H-coal process.
- From 1965 to 1968, $1.75 million was spent.
- From 1968 through 1970, approximately $1.5 million was spent.
- From 1971 through 1973, $5 million (contributed by private industry and HRI) was spent.

Related Pilot Plant Work. The results through 1974 demonstrated that H-Coal was technically feasible and had strong commercial potential. To realize this potential, ERDA (now DOE) and industrial partners initiated a pilot plant program for processing 600 ton/day (545 metric ton/day) of coal to produce 2000 bbl/day of low-sulfur fuel oil, or for

MAJOR LIQUEFACTION PROCESSES

processing 250 ton/day (227 metric ton/day) of coal to produce 650 bbl/day of synthetic crude oil.

The primary objective of the program is to build and operate the pilot plant to provide final yield confirmation and to serve as an engineering data source and an equipment testing facility as a last step prior to commercialization. In 1976 a contract was signed with Ashland Synthetic Fuels, Inc. (ASFI) for the construction and operation of this plant at Catlettsburg, KY, adjacent to the Ashland Oil Refinery. This was a joint government-industry program, with industry represented by ASFI, Standard Oil Company (Indiana), Mobil Oil Corporation, Continental Oil Company, Atlantic Richfield Oil Company, Sun Oil Company, Shell Oil Company and the Electric Power Research Institute. Government sponsorship was obtained through ERDA (DOE) and the Commonwealth of Kentucky. When it began operation in 1980, the Catlettsburg H-Coal pilot plant was the largest coal conversion facility in the United States.

Current Status

The H-Coal process is currently at a stage in its development that makes it ready for the initial phase of a full-scale commercialization program. In 1980, the 250- to 600-ton/day (227 to 545 metric ton/day) pilot plant at Catlettsburg, KY, will come on-stream to provide equipment design confirmation. This will be the last step in the commercialization of the process that started at the 25-lb/day (11.4 kg/day) level in 1964. Work was begun on the 3-ton/day (2.7 metric ton/day) PDU in 1967 and, since 1974, this facility has been involved in a program in which key items, process and operating variables, have been demonstrated, including:

- an ebullated-bed reactor that has been scaled to commercial design;
- continuous recycle of a stream containing solids and residuum;
- continuous production of distillate recycle streams maintaining solvent balance and continuous distillate production during runs of up to 30 days;
- recycle hydrogen produced in the same manner as that envisioned for a commercial plant;
- continuous catalyst addition and withdrawal with achievement of equilibrium activity yields; and
- direct-fired slurry charge heater.

In addition to normal operation, the PDU has been subjected to the same type of operating upsets that would be expected in a commer-

46 LIQUEFACTION

cial unit. The experience gained thereby has led to the development of operating techniques and a sparing policy that permits the operation to be continued without a total shutdown. The operating upsets have included:

- short-term power failure,
- compressor loss, and
- pump loss.

In addition to PDU experience, the H-Oil unit in Kuwait has achieved runs of almost 2-year duration prior to planned shutdowns.

While the commercial design is being performed, the Catlettsburg pilot plant will be operational and will provide information that can be included in the final design and equipment selection. The purpose of the pilot plant is not to prove process operability (this has been done at the 3-ton/day (2.7 metric ton/day) PDU level), but to provide an engineering data source, an equipment testing facility, and a confirmation of product yield. The size selected was large enough to utilize equipment of commercial size.

The 250- to 600-ton/day (227- to 545-metric-ton/day) pilot-plant feed rate is rather small compared with the 20,000-ton/day (18,000 metric ton/day) feed rate of a commercial plant. However, the commercial plant will have parallel reactors and associated equipment not much larger than those found in the pilot plant. Scale factors must be compared relative to the rates anticipated in a single train or in a single pass in a train. Comparable sizes of equipment for the H-Coal pilot plant, the commercial-size H-Coal facility, and the Kuwait H-Oil unit are given in Table 2-10.

The above H-Coal scaleup factors on equipment are within sound engineering practice. Downstream equipment, such as exchangers and

Table 2-10. Equipment Sizes for H-Coal and H-Oil Facilities[a]

Item	H-Coal Pilot Plant	H-Coal Commercial	H-Oil Kuwait	H-Coal Scale Factor
Reactor	5 ft	12 ft/train	13.5 ft	2.5
	1.5 m	3.66 m/train	4 m	
Heater Tubes	4.5 in. o.d.	5 in./pass	5 in./pass	
	11.4 cm o.d.	12.7 cm/pass	12.4 cm/pass	None
Charge Pumps	100 gal/min	500 gal/min	1,200 gal/min	
	378 liter/min	1892 liter/min	4542 liter/min	5
Hydroclones	10-mm orifice each	10-mm orifice each		None

[a]Source: H-Coal Technical Proposal, 1978.

MAJOR LIQUEFACTION PROCESSES 47

columns, are of conventional design. The sizes of the operational Kuwait H-Oil unit and the planned H-Coal commercial unit are nearly identical.

The pilot plant will provide the data required for the final selection of materials of construction and for the determination of routine maintenance requirements. These data will be available when it is necessary to make final decisions on materials, spare parts and sparing policy.

The H-Coal process is at a stage where engineering on a full-scale commercial plant can proceed concurrently with pilot-plant operation. Detailed equipment specifications needed for procurement will not be required until the pilot plant is in operation. Several promising new catalysts with improved denitrogenation activity have been evaluated on an experimental level. These catalysts are improved formulations impregnated onto slightly modified commercial supports. An improved catalyst could be in commercial production in time for the initial loading of a commercial plant.

HRI has made a commitment that the H-Coal technology will be made available through license to any responsible party at royalty terms previously established and identified with the DOE.

The 250- to 600-ton/day (227- to 545-metric ton/day) Pilot Plant

The H-Coal pilot-plant project is being conducted in three phases:

- Phase I—PDU confirmation and pilot plant design,
- Phase II—construction, and
- Phase III—pilot plant operations.

Phase I:

Phase I had two objectives:

1. PDU confirmation of key design features, which involves: demonstration of reliable operation and resolution of any problems revealed in the PDU or engineering programs
2. detailed engineering design of the pilot plant to operate in both the syncrude and boiler fuel modes, followed by procurement of long- and short-lead-time equipment.

Phase I has been completed, with the following results:

48 LIQUEFACTION

- an operational demonstration in the boiler fuel mode (this was an uninterrupted run to a catalyst life of 3000 lb coal/lb catalyst 3000 kg coal/kg catalyst);
- an operational demonstration of the boiler fuel mode at the high exit gas velocity that will occur in the pilot plant;
- a demonstration of the boiler fuel mode, with high residuum concentration in the reactor;
- a demonstration of the syncrude mode, with high residuum concentration in the reactor;
- utilization of an on-line, continuous hydroclone system similar to the pilot plant design to achieve a high residuum concentration in the reactor (this system was used in all PDU operations);
- demonstration of solid-liquid separation in the syncrude mode by vacuum distillation to produce a flowable material with up to 55% solids;
- demonstration in the syncrude and boiler fuel modes of an on-line antisolvent precipitation and settling system for solid-liquid separation; and
- demonstration of improved reactor internals.

The objectives of the PDU program were attained, and the operability of the H-Coal process was clearly demonstrated. HRI concluded that the process was feasible for scale-up to pilot plant size, the design basis originally proposed for the pilot plant was adequate and no modifications were required. At the same time that the laboratory program was conducted, detailed engineering and equipment procurement studies were completed. Environmental impact statements were prepared and permit applications were filed.

Phase II

The major task of phase II is the construction of the pilot plant, which was completed in the summer of 1980. The cost of construction was $135 million. Concurrent experimental work by HRI was also being performed to assure successful operations of the pilot plant with a minimum of downtime. The cost of the research and engineering effort is $35 million for PDU, bench-scale and cold-flow model work.

The objectives of the PDU program are:

- to simulate the pilot plant design as closely as possible by: testing the hot slurry feed system, demonstrating catalyst addition and withdrawal, and testing the two-stage pressure reduction system;

MAJOR LIQUEFACTION PROCESSES 49

- to duplicate the operating conditions for the pilot plant as closely as possible;
- to develop pilot-plant emergency operating procedures by exploring potential failure modes and responses on the PDU;
- to provide training for pilot plant personnel;
- to test an antisolvent precipitation and settling system and to test for possible antisolvent degradation.

The bench-scale program will evaluate:

1. the use of alternative catalysts for improved conversions;
2. the use of improved denitrogenation catalyst;
3. lower pressure operation; and
4. the yield structure obtained from processing Kentucky coal.

Throughout the entire period of H-Oil and H-Coal development, a 6-in. (15-cm) diameter, low-pressure, ambient-temperature reactor model using kerosene and nitrogen was employed to diagnose hydraulic problems and test solutions to them. During the phase I H-Coal program, the model was used to study the problem of gas entrainment in the recycle liquid and to test revised recycle pump suction cup designs. The design that worked best in the cold model was subsequently installed in the PDU and permitted operation at the highest gas velocity ever achieved. Because of the proven success of this technique on the PDU scale, it was decided to test the scale-up of the revised reactor internals from the PDU to pilot-plant scale by building a full-size model of the pilot plant reactor and internals. When the redesigned reactor internals were tested using a kerosene-nitrogen system, they showed the same degree of improvement over the prior design as had been shown in the smaller scale tests. As a result, the revised design was approved for installation in the pilot plant.

Phase III

The two-year program planned for Phase III is shown in Table 2-11. The cost of operations is $125 million. After three months of plant shakedown and three months of break-in operation, there will be six months of operation in each of the following modes: the syncrude mode, the boiler fuel mode and an intermediate mode. The expected yields from Illinois #6 coal in the boiler and intermediate modes are given in Tables 2-12 and 2-13. A coal analysis is given in Table 2-14 and the PDU results are given in Table 2-15. The pilot plant operations will terminate in the spring of 1982.

Commercial H-Coal Liquefaction Plant

The conceptual design of a commercial H-Coal plant is based on a fully integrated facility that includes provisions for the receipt of raw materials, storage, coal cleaning and preparation, the conversion of coal into liquid and gaseous hydrocarbons, separation and purification of the hydrocarbon products, and the handling, storage and shipment of the plant products. All required plant support facilities and utilities are included. The plant will process approximately 20,000 ton/day (16,482 metric ton/day) of high-sulfur (4%) Illinois #6 coal. The plant will

Table 2-11. Pilot-Plant Phase III Technical Management Plan

Year	Duration (mo)	Space Velocity lb/hr/ft³	kg/hr/m³	Operation Phase	
1	3			Plant shakedown	Oil (no coal)
	3	31	0.4	Break-in operation	Kentucky
	3	31	0.4	Syncrude mode	Illinois
	3	31	0.4	Syncrude mode	Wyodak
2	3	0–78	0–1	Startup in boiler fuel mode	Kentucky
	3	78	1	Boiler fuel mode	Illinois
	3	60	0.78	Intermediate mode	Kentucky
	3	45	0.6	Intermediate mode	Illinois

Table 2-12. Projected Pilot-Plant Yields: Boiler Mode[a]

Reactor Yields	lb/100 lb MF Coal	°API	B/D	%S
C_1-C_3	6.36	113.0	63.7	0.07
C_4	1.26	47.0	515.1	0.09
C_5-400°F	13.37	21.7	377.0	0.27
400–650°F	11.44	−0.1	318.6	0.86
650–975°F	11.27	−20.0	770.0	
975°F+	32.09			
Unconverted Coal	7.14			
Ash	10.00			
H_2O	7.01			
$CO + CO_2$	0.83			
NH_3	0.72			
H_2S	2.39			
Total	103.88		2044.4	
Sulfur in 400°F+ Oil (wt %)				0.58

[a]Space velocity = 78 lb coal/hr/ft³ reactor; coal throughput = 531 ton/day; hydrogen partial pressure = 1650 psi; catalyst replacement rate = 1.0 lb catalyst/ton coal.

produce a nominal 50,000 bbl/day of liquid hydrocarbons, 570.5 ton/day (long tons) (580 metric ton/day) of sulfur, 119.4 ton/day (123 metric ton/day) of anhydrous ammonia and 29.7×10^6 scf/day (0.84×10^6 m³/day) of high-Btu [1070 Btu/ft³ (39,911 J/1)] gas. The products are shown in Table 2-16; the overall thermal efficiency of the

Table 2-13. Projected Pilot-Plant Yields: Intermediate Mode[a]

Reactor Yields	lb/100 lb MF Coal	°API	B/D	%S
C_1-C_3	8.55	113.0	49.0	0.06
C_4	1.67	47.0	394.7	0.08
C_5-400°F	17.73	23.1	309.1	0.24
400–650°F	16.09	0.3	180.5	0.47
650–975°F	11.05	−20.0	309.1	
975°F+	22.31			
Unconverted Coal	5.39			
Ash	10.00			
H_2O	7.81			
$CO + CO_2$	0.85			
NH_3	0.92			
H_2S	2.62			
Total	104.99		1242.4	
Sulfur in 400°F+ Oil (wt %)				0.29

[a]Space velocity = 45 lb coal/hr/ft³ reactor; coal throughput = 306 ton/day; hydrogen partial pressure = 1650 psi; catalyst replacement rate = 1.0 lb catalyst/ton coal.

Table 2-14. Analysis of Coal

	Wyoming Subbituminous Wyodak	Illinois #6 Burning Star
Proximate Analysis, Dry Basis (wt %)		
Ash	8.28	11.13
Volatile Matter	42.26	36.01
Fixed Carbon	45.46	52.86
Ultimate Analysis, Dry Basis (wt %)		
Carbon	67.08	70.83
Hydrogen	4.91	5.04
Nitrogen	0.67	1.45
Sulfur	0.87	4.06
Ash	8.28	11.13
Oxygen	18.19	7.42
Heating Value (Btu/lb)	11,552	12,620
(kcal/kg)	6,411	7,022

[a]Source: Phase II Laboratory Support, March 1978.

52 LIQUEFACTION

plant is shown in Table 2-17. A schematic diagram of the commercial H-Coal plant is given in Figure 2-5.

Coal is received at the plant by rail or by barge, and is conveyed to a live storage pile. A 60-day inventory of coal is maintained in dead

Table 2-15. Typical H-Coal Process Yields, 3-ton/day Process Development Unit, Continuous Catalyst Replacement[a]

	Wyodak Coal	Illinois #6 Burning Star Mine
Yields (wt %) of dry coal		
C_1-C_3	12.35	12.4
C_4-400°F	25.82	18.5
400–650°F	10.85	20.65
650–975°F	7.74	4.85
975°F+ Liquid	11.63	20.71
Unconverted Coal	8.50	4.88
Ash	7.80	11.65
Water	16.97	7.22
$CO + CO_2$	2.98	0.59
$NH_3 + H_2S$	1.61	3.70
Total Yield	106.25	105.14
Hydrogen Requirement	6.25	5.14
Total of 975°F+ Liquid, Unconverted Coal and Ash Yield	27.93	37.23
Actual Vacuum Still Bottoms Yield in PDU Operation	33.42	42.55

[a]Source: Phase II Laboratory Support, March 1978.

Table 2-16. H-Coal Commercial Plant Design Material and Product Balance, Stream-Day Basis[a]

Input	
Coal (tons)	20,000
(metric tons)	19,685
Output Product	
Middle Distillate (bbl)	27,940
High-Octane Reformate (bbl)	15,260
Butane (bbl)	3,290
Propane LPG (bbl)	3,510
Sulfur (long tons)	570.5
(metric tons)	580
Ammonia (long tons)	119.4
(metric tons)	123
Synthetic Pipeline Gas (10^6 scf)	29.7
(10^6 m^3)	0.84

[a]Source: H-Coal Technical Proposal, September 1978.

MAJOR LIQUEFACTION PROCESSES 53

storage. Coal from the live storage pile is conveyed to the coal cleaning and drying plant, where the run of mine coal is crushed to a nominal minus 3 in. (minus 7.6 cm). This crushed coal is fed to a dense-media cleaning plant and processed to produce a clean coal containing no more than 8 wt% ash. The middlings fraction is sent to the steam boiler plant and used as fuel; excess middlings are sold. The reject fraction, containing primarily ash, is sent to settling ponds. The clean coal is conveyed to storage bins which feed fluid-bed dryers, where the moisture content of the coal is reduced to about 2 wt%. The dry coal is fed to a closed-loop system for fine grinding to a size of minus 100 mesh. The fine coal is pneumatically conveyed to the feed hoppers in the slurry-preparation system of the H-Coal plant.

The H-Coal reactor is a critical element in the H-Coal process. In the H-Coal process, coal is slurried with a process-derived oil and introduced into an ebullated reactor containing a catalyst. Hydrogen reacts with the coal to produce liquid and gaseous products. An H-Coal reactor was illustrated schematically in Figure 2-4. For purposes of clarity, ebullating flow taken from the external separator is shown. Normally, an internal recycle cup is used to separate the recycle (ebullating) from the reactants. Coal-oil slurry and hydrogen are introduced into the bottom of the reactor, as is the controlled-reactor recycle stream. The cross-

Table 2-17. Overall Thermal Efficiency of Commercial H-Coal Plant

	Input (10^6 Btu/day)	Output (10^6 Btu/day)	10^6 kWh/day	% of Total
Feed				
Coal fed to hydrogenation	373,177		109.27	
Coal to steam plant	39,310		11.51	
Coal for drying clean coal	3,703		1.08	
Coal lost in refuse	31,327		9.17	
Product				
Gas oil		176,025	51.54	53.5
High-octane reformate		85,921	25.16	26.1
Pipeline gas		31,883	9.34	9.7
n-Butane		14,259	4.55	4.3
Propane		13,544	4.32	4.1
Ammonia		2,309	0.74	0.7
Sulfur		5,082	1.49	1.6

HHV of Coal = 11,167 Btu/lb (6,204 cal/kg)

Efficiency = (329,023/447,650) × 100 = 75%

[a]Source: H-Coal Technical Proposal, September 1978.

54 LIQUEFACTION

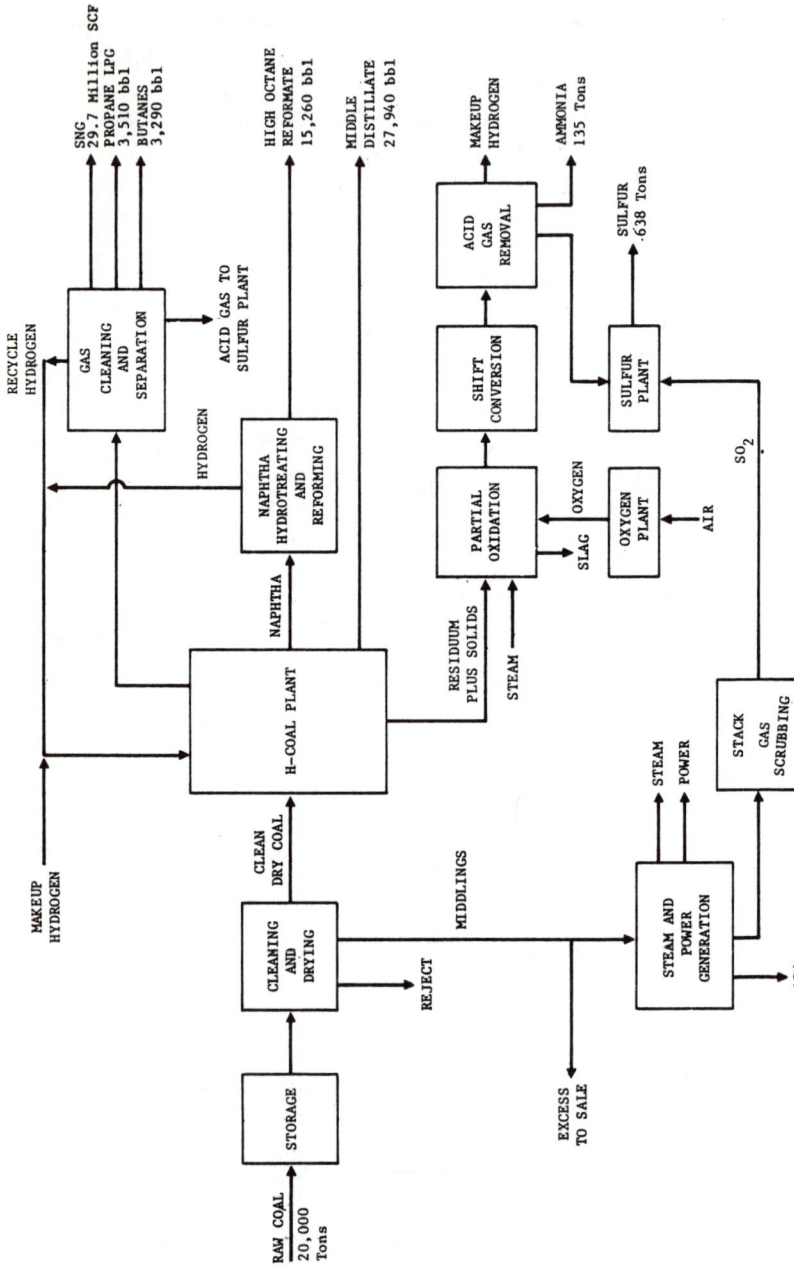

Figure 2-5. Commercial H-Coal complex block flow diagram (source: H-Coal Technical Proposal, 1978).

hatched area in the figure indicates the catalyst bed prior to expansion. The amount of bed expansion is controlled by the physical properties of the feed and recycle and by the upward flowrate. The area with diagonal lines indicates the amount of expansion. Catalyst particles within the expanded bed are in constant, random motion. There is a gradient of catalyst particles from the bottom to the top of the bed, with the heavier, less buoyant particles tending to collect near the bottom. A very sharp interface between the top of the ebullating catalyst bed and the oil slurry is obtained.

The effluent is separated into vapor and liquid streams. Recycle liquid is pumped back into the reactor to control the rate of liquid upflow in the reactor and the amount of catalyst bed expansion. Because the catalyst is constantly in motion, a portion of the catalyst can be withdrawn and replaced with fresh catalyst. Additionally, because the catalyst is totally mixed, the average age (or activity) is controllable by the amount of addition and withdrawal. This is done as needed, with the amount of catalyst charged normally being 1–3% of the inventory. A further consequence of the catalyst motion is that the reactor is nearly isothermal and, as a result, the temperature can be controlled closely. Since the hydrogenation reaction is exothermic, the reactor temperature is controlled by introduction of the feed and/or recycle hydrogen at temperatures somewhat lower than the reaction temperature. Thus, the reactor serves as an efficient heat exchanger, thereby minimizing the capital cost for heating and cooling equipment and eliminating a portion of the thermal inefficiencies caused by heating the feed and cooling the reactor effluent.

Plant Design

The clean, dried coal is mixed with a processderived recycle oil to produce a slurry, which is pumped to reactor pressure and mixed with a part of the hydrogen recycle stream. This stream is heated and enters the reactor at a temperature lower than the reactor temperature. The balance of the hydrogen is heated separately and fed to the reactor. The reaction takes place in the presence of a commercially available catalyst. This conceptual plant design includes eight H-Coal reactors. Products generated from the reaction are withdrawn from the reactor at the rate at which they are produced, along with the recycled streams, and the mixture is separated into vapor and liquid streams.

The vapor stream is cooled and processed to recover light products and unreacted hydrogen. The unreacted hydrogen is recycled, and the remaining materials are sent to further processing.

56 LIQUEFACTION

The liquid is depressured and flashed, and the vapors are cooled and sent to the fractionation system. The liquid from the flash is separated into a high-solids-content stream and a low-solids-content stream in hydroclones. The low-solids-content stream is returned to the slurry preparation system as part of the slurry oil. The high-solids-content stream flows to an atmospheric distillation tower. The overhead from the atmospheric tower is sent to the fractionator. The bottoms from the atmospheric tower flow is sent to a vacuum distillation tower. The overhead from the vacuum distillation tower is also sent to the fractionator. The bottoms from the vacuum tower containing heavy hydrocarbon liquids, plus solids which consist of unconverted coal and ash, are used as feed to the hydrogen plant.

All of the accumulated solids-free liquids from the H-Coal plant are fed to a fractionator, where they are divided into light gases, a naphtha fraction, middle distillate and heavy oil. All of the heavy oil and a small amount of middle distillate are recycled to the slurry preparation section. The remainder of the middle distillate fraction is a final plant product. The naphtha fraction is sent to the naphtha hydrotreating and reforming plant for further processing. The overhead gases from the fractionator, together with previously flashed light hydrocarbon gases, are sent to the gas cleaning and separation section for further processing.

The makeup hydrogen requirements for the H-Coal complex are obtained by the partial oxidation of the vacuum tower bottoms, which contain heavy hydrocarbon liquids, unconverted coal and ash. The vacuum tower bottoms, steam and oxygen are fed to an entrained, slagging, partial oxidation unit to produce a synthesis gas containing primarily hydrogen and carbon monoxide.

After particulate removal, the synthesis gas is mixed with the steam and fed to a shift conversion unit, where hydrogen is produced by the water-gas reaction. The gases after shift conversion are sent to the acid gas removal section to remove carbon dioxide and hydrogen sulfide. The resulting clean gas is the makeup hydrogen stream for the H-Coal plant. Acid gases are sent to the sulfur plant for sulfur recovery. A conventional air separation plant is used to produce the oxygen required for the partial oxidation of the vacuum tower bottoms. For this conceptual plant, two 1800-ton/day (1634-metric-ton/day) oxygen plants are required.

Stabilized naphtha, a nominal C_5-400°F (204°C) fraction, is fed to a naphtha hydrotreating section, mixed with hydrogen produced in the reforming section, and fed to the hydrotreating reactor. This reactor uses a commercial nickel-molybdenum on alumina catalyst. The primary function of the naphtha hydrotreater is to reduce and/or remove the

MAJOR LIQUEFACTION PROCESSES 57

nitrogen, oxygen and sulfur compounds from the naphtha. This hydrotreater is operated at conditions more severe than those normally existing in conventional petroleum naphtha hydrotreaters because of the higher nitrogen, oxygen and sulfur contents in naphtha derived from H-Coal. The hydrotreater effluent is cooled and fed to a high-pressure flash drum for the recovery of hydrogen. A portion of the hydrogen is recycled to the hydrotreater, and the remainder is compressed and recycled to the H-Coal section. Liquid from the high-pressure flash drum is partially depressured in a low-pressure flash drum. The offgases from the low-pressure flash drum are sent to the gas plant. Liquids from the low-pressure flash drum are sent to a stabilizer to prepare a C_5-400°F (204°C) naphtha stream, which is dried and sent to the reforming reactors.

The dried reformer naphtha feed is mixed with recycled hydrogen obtained from the reformer, high-pressure flash drum, heated and fed to the first reforming reactor. The naphtha feed is very highly naphthenic. Since the conversion of naphthenes to aromatics is highly endothermic, multiple reactors in series with interstage heaters are required. Four reactors are used in series. Because of the naphthenic nature of the feed, little or no dehydrocyclization and/or isomerization are required to produce a very high-octane reformate. The relatively mild conditions required permits the use of conventional platinum on alumina reforming catalyst to take advantage of its relative tolerance to sulfur and nitrogen compounds. The stabilized reformate produced is in excess of 100 research octane and is a primary plant product.

The gas streams from the H-Coal section and the naphtha hydrotreating and reforming section are mixed and compressed for feed to a light-oil washing system. The light-oil wash removes all of the naphtha components and most of the butanes. Lighter gases are sent to an acid gas removal system for the removal of carbon dioxide and hydrogen sulfide. Clean gases from the acid removal system are sent to molecular sieve treaters to remove traces of carbon dioxide, hydrogen sulfide and moisture before being fed to a cryogenic cold box. The cryogenic cold box separates the lighter gases into a relatively high-purity hydrogen stream, a synthetic natural gas stream, and a propane-butane mixture. The hydrogen stream is recycled for use in the H-Coal section. The synthetic natural gas stream is compressed to pipeline pressure and sold as a plant product. The propane-butane mixture from the cryogenic unit is combined with the propane-butane mixture from the light oil wash and separated into propane and butanes in a depropanizer. The propane liquified petroleum gas and mixed butanes are plant products. The naphtha recovered in the light oil wash is sent to the naphtha hydrotreating

58 LIQUEFACTION

and reforming section for further processing. The acid gases from the acid gas removal section are sent to the sulfur plant for sulfur recovery.

Three 50% steam boilers and three 50% turbogenerators are provided to produce the steam and power requirements for the complex. All three steam boilers are fueled by the middlings fraction of the coal cleaning plant. A stack gas scrubbing system is used to remove sulfur dioxide from the flue gases. Three 50% absorbers are included. The sulfur dioxide from the stack gas scrubbing system is sent to the sulfur plant for sulfur recovery. Tail gases from the sulfur are returned to the boiler stack gas scrubbing system for final cleanup. The sour water from various sections of the complex is collected and processed for recovery of ammonia.

Economic Analysis

A preliminary economic analysis of a 50,000-bbl/day H-Coal commercial plant has been performed. A more extensive analysis is being performed as part of the commercial plant design now in progress. The capital requirement was estimated at $1 billion (1978 dollars). The operating costs were estimated on the basis of specific coal, labor, payroll overhead and miscellaneous expenses. A discounted cashflow rate of return analysis was performed using criteria on investment tax credits, depreciation, loss carryover, three-year reduced capacity startup, and by-product credits for sulfur, ammonia and high-Btu gas. Inflation rates were projected for oil, coal, construction and general expenses, and the initial average selling price for the liquid products in 1978-value dollars necessary to finance the project was computed. For coal prices varying between $20 and $30/ton and rates of return from 9 to 15 percent, the product prices ranged from $12.60 to $21.10/bbl. The estimated value of the liquid products from the H-Coal plant is slightly higher than comparable petroleum products because of higher aromatic content and lower values of sulfur, oxygen and nitrogen. A selling price of $17/bbl, in July 1978 value, is estimated for the H-Coal plant liquid products.

The product liquids from an H-Coal plant consist of more than half middle-distillate fuel oil and over a fourth high-octane reformed naphtha gasoline-blending stock. The remainder is butane and propane liquefied petroleum gases. Synthetic pipeline gas constitutes almost one-tenth of the energy output of the plant. Demand for the fuel oil and naphtha is forecast to remain strong. The liquefied petroleum gas and synthetic pipeline gas should be marketable, but at reduced prices. A second

MAJOR LIQUEFACTION PROCESSES 59

analysis, by Kunesch et al., considered site-specific effects on liquid product costs. Two key factors in the economic analysis were the means of hydrogen generation and whether the facility manufactures its own electrical power or purchases it. Hydrogen may be produced by the accepted procedure of steam reforming of the light gases made in the liquefaction step. Another alternative is the partial oxidation of the mixture of ash, unconverted coal and residuum which comes from the H-Coal vacuum distillation bottoms, a process under development.

Assuming that a customer is available for the fuel gas produced from partial oxidation of sufficient vacuum bottoms to place the plant into hydrogen balance, a revenue of $2.50/10^6$ Btu is obtained. If the fuel gas cannot be sold, then the bottoms are carbonized and the coke used for electric power generation; hydrogen is produced by steam reforming of light hydrocarbons.

The economic analysis for the partial oxidation of bottoms was performed, assuming 55% debt at 8% interest and 45% capital. A 10% discounted cashflow return on equity was used and naphtha product value was set equal to fuel oil. The breakdown of the oil selling price in 1976-value dollars was computed to be:

Coal at $15/ton	$ 6.74
River water	.12
Catalyst and chemicals	.74
Labor, supervision and overhead	.66
Maintenance	1.80
Insurance and taxes	1.57
Total operating cost	$11.63
Capital-related expense	9.95
By-product credit	−3.51
Total oil selling price	$18.07

Additional economic factors investigated in the study include the sensitivity of the process options to the percent of capital investment, the cost of purchased electric power, coal cost, selling prices for naphtha and fuel oil, and the marketability of the fuel gas produced in the partial oxidation of the bottoms. The effect of these parameters depends on the specific site chosen for the location of a commercial plant and the market conditions prevailing at the startup and during the 20-year lifetime of the plant. The initiation of commercial plant construction will not begin until issues such as these are resolved. Additional issues that have a significant effect on process economics are interest rates, which determine over half the production costs, and taxation policy.

Assessment

Scale-up of the H-Coal reactors is facilitated by the experience of the developer with larger H-Oil reactors in commercial service. The H-Coal flexibility in controlling processing time in the reactor is good. Acceptable catalyst lifetime must be demonstrated with different types of coal. The optimum disposition of the vacuum tower bottoms has not been determined. Pilot plant operation will evaluate solvent de-ashing methods. Use of the bottoms as feed for a commercial gasifier or as a boiler fuel are possibilities. Predicted product costs are low and will have to be revised upward. The fact that the Catlettsburg pilot plant construction has encountered cost increases and schedule delays supports the assumption that costs estimates are low. Successful pilot plant operation will be required to sustain confidence in the scale-up of the process to commercial size.

SOLVENT-REFINED COAL

Process Description

The solvent-refined coal (SRC) process separates the combustible organic material in the coal from the sulfur, nitrogen and ash to produce a higher heat content fuel that can be burned with minimum environmental treatment. Heat and pressure, combined with the catalytic activity of minerals contained in the coal ash, are used to liquefy and desulfur the coal without the extra complicated steps of catalytic processes. Pulverized coal is slurried with a recycled solvent having hydrogen donor properties as shown in Figure 2-6. Hydrogen gas is added to the mixture, which is heated to 800°F (427°C) and fed into a dissolver operating at 1400–2500 psi (98.4–176 kg/cm^2). More than 90% of the coal is dissolved in 30–60 min residence time in the preheater and dissolver. The degree of dissolution depends on the reactivity of the coal and competition from other reactions that depolymerize and hydrogenate the coal, hydrocrack the solvent to lower-molecular-weight hydrocarbons, and remove sulfur and oxygen by combination with hydrogen.

The dissolver product stream is depressurized and separated into its components, consisting of light hydrocarbon gases, excess hydrogen, dissolved coal, unreacted coal and undissolved ash. The gases are scrubbed to remove hydrogen sulfide and carbon dioxide. Some of the hydrogen

MAJOR LIQUEFACTION PROCESSES   61

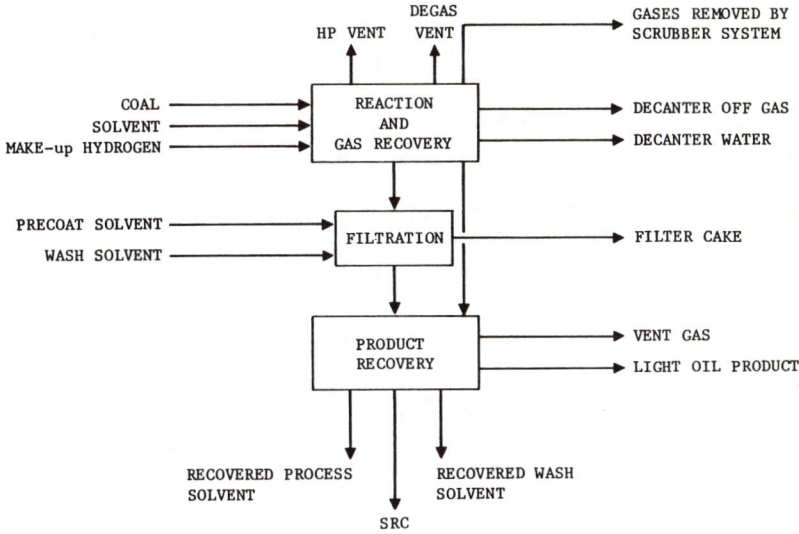

Figure 2-6.   SRC process flowsheet (source: Quarterly Technical Progress Report, 1977).

is separated and recycled to the preheater, while the rest forms a medium-Btu fuel gas for sale or plant process use. Undissolved coal and mineral matter are separated from the SRC solution by filtration. The filtered liquid is distilled to separate the 550–800°F (288–427°C) boiling point solvent from the SRC product which is solidified by cooling. Light organic liquid and hydrocarbon gases are separated and recovered as by-products.

The advantages claimed for the SRC process are that more than 90% of the feed coal is processed into some form of usable fuel, with minimum consumption of hydrogen. The product can be a solid bottoms fraction from distillation, with a solidification temperature of 350°F (177°C), or hydrocracked and hydrotreated to produce heavy oils that are liquids at ambient temperatures. The treatment severity controls the amount of sulfur, oxygen and nitrogen remaining in the product. Higher conversions reduce the concentrations of sulfur, nitrogen and oxygen, but increase the hydrogen consumption and residence time in the reactor.

No expensive catalysts or treatment steps that need additional equipment to handle catalysts are required. For successful operation and high productivity, the SRC process requires a reactive bituminous coal. Several bituminous coals have successfully been processed. The yield of SRC can be varied over a large range by varying process conditions.

62 LIQUEFACTION

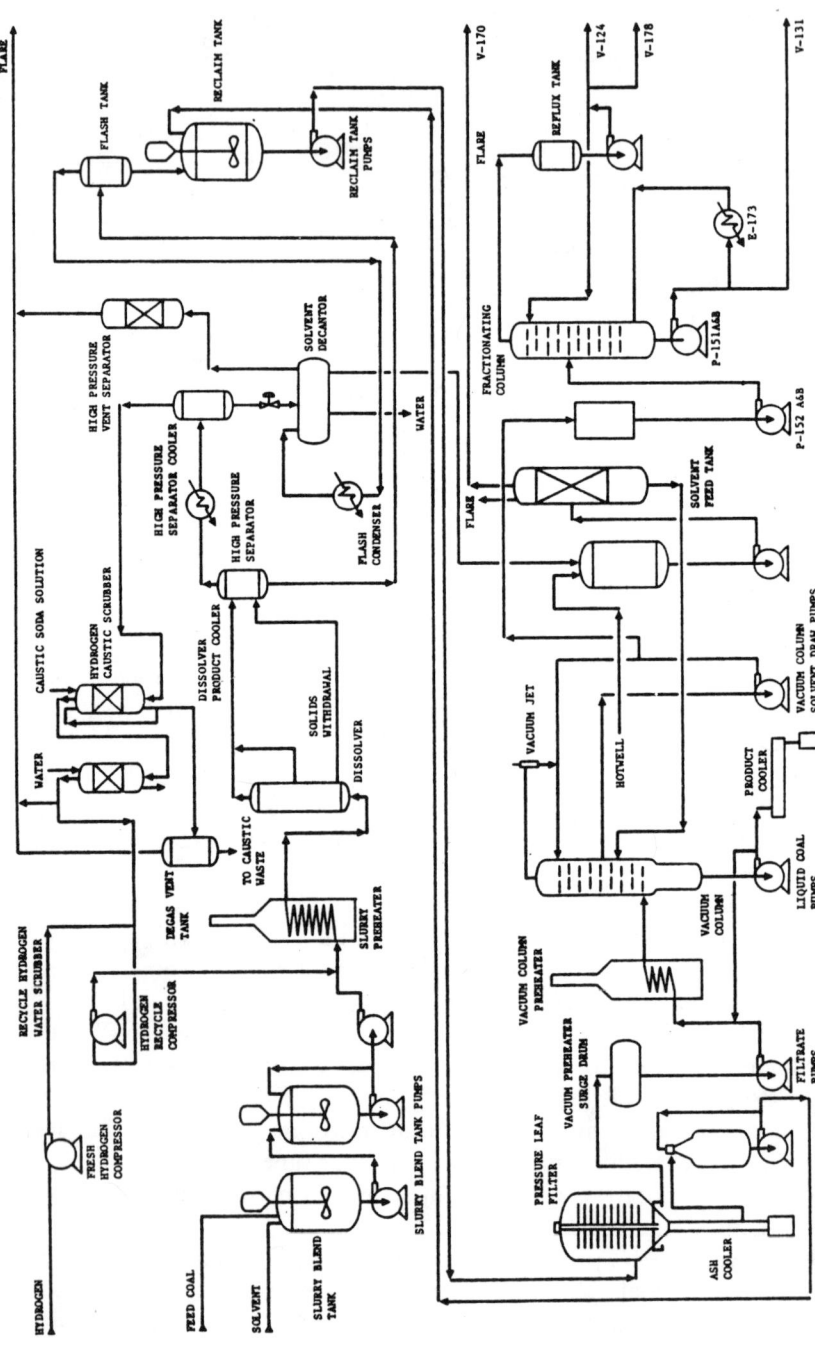

Figure 2-7. Flowsheet for Wilsonville SRC pilot plant (source: Quarterly Technical Progress Report, 1977).

Thermal efficiencies are maximized at 67–70% when the solvent-refined coal yield is between 15 and 45% of the dry coal feed rate.

SRC-I Process Description

The Wilsonville, AL, SRC pilot plant illustrated in Figure 2-7 pulverizes coal to minus 200 mesh and mixes it with a process-generated solvent in a slurry blend tank. The boiling range of the solvent is 450–900°F (232–482°C). A hydrogen-rich gas may be added to the coal slurry before it enters the preheater, or hydrogen injection may occur between the preheater and the dissolver. The feed gas stream consists of scrubbed recycle gas plus sufficient makeup hydrogen to bring the overall gas composition to 85% hydrogen by volume. The coal slurry and gas are heated in a 600-foot-long (183 m) helical coil of 1.25-in. (3.2 cm) stainless steel pipe by an oil burner located at the bottom of the slurry preheater. The slurry leaving the preheater passes into a 1-ft (0.3 m) diameter, 23-ft (7 m) high dissolver that is centrifugally cast of high-chrome stainless steel. It can be operated at 800–875°F (427–468°C), 1400–2500 psi (98.4–176 kg/cm^2), and can provide residence times of 0.5–1.5 hr.

Because the reactions (hydrogenation of the solvent and coal, hydrocracking of the solvent and coal into lower-molecular-weight compounds, formation of water) occurring in the dissolver are exothermic, the slurry increases in temperature over the value at the preheater discharge. A solids withdrawal system may be operated intermittently or continuously to control the dissolved solids accumulation.

Effluent from the dissolver is cooled to 500–650°F (260–343°C) and separated into vapor and slurry phases by the high-pressure separator. Vapor from the separator is cooled to 150°F (65°C) by a high-pressure cooler and is then passed into a high-pressure vent separator. The water and organic compounds are fed to a light-solvent recovery column. The vapors include unreacted hydrogen, light hydrocarbon gases, hydrogen sulfide and carbon oxides.

The slurry is depressurized through high-pressure letdown valves into a flash tank, where vapors are separated. The slurry now proceeds to the most critical phase of the SRC-I process (Figure 2-8). Filtration of the SRC product is performed to separate the high-melting-point hydrocarbon from the ash and unreacted coal, which is in the form of micron-sized solid particles. Filter operations have proven to be the source of numerous mechanical problems and many plant shutdowns. A pressure-leaf filter is used with metal wire screens, having a filtration area of 100 ft^2 (9.3 m^2). The filter is operated at 480–580°F (249–304°C) and

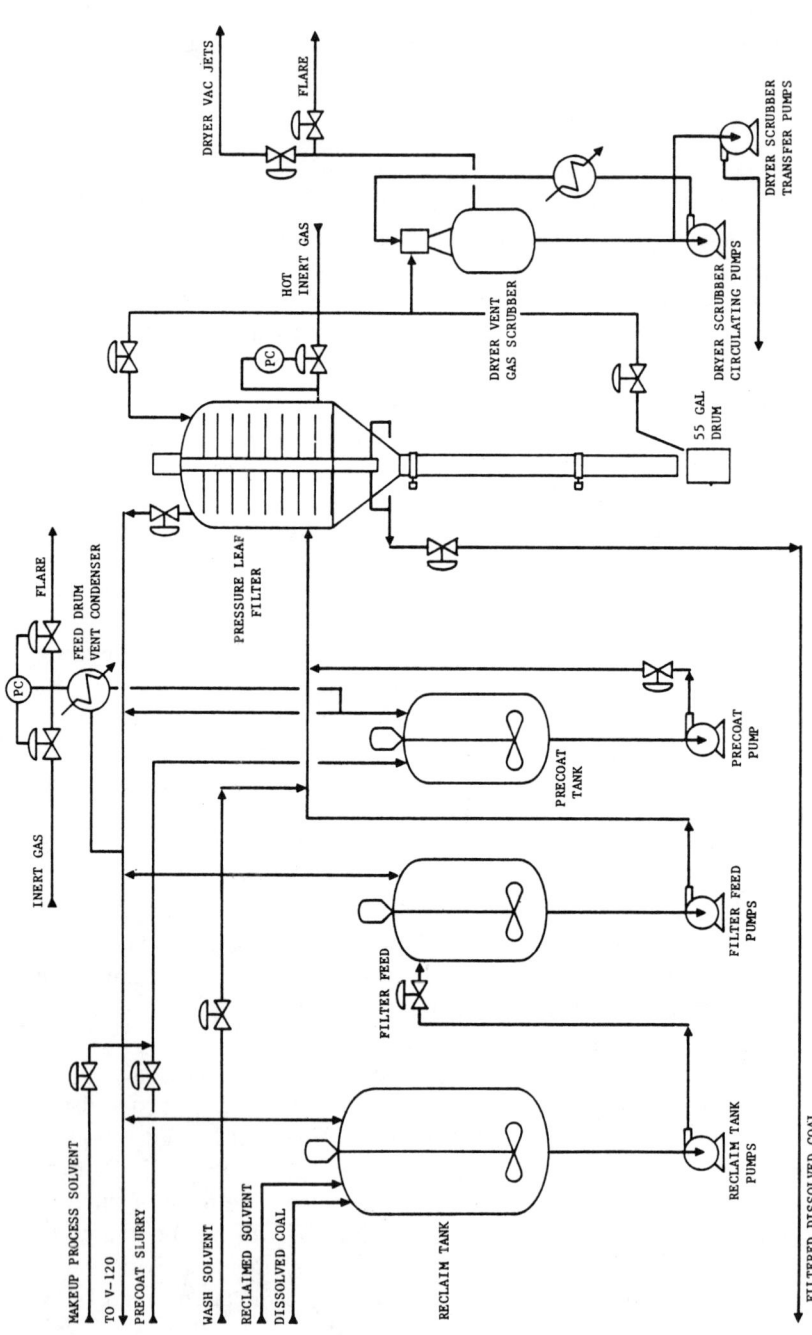

Figure 2-8. Filtration flowsheet (source: Quarterly Technical Progress Report, 1977).

150–200 psi (10.5–14 kg/cm^2), with a maximum pressure drop of 80 psi (5.6 kg/cm^2) between the slurry inlet and filtrate outlet. An automatic sequence of operations is performed:

1. precoating the filter with a high-boiling-point process solvent;
2. filtration of the slurry;
3. washing of the filter with a 350–450° (177–232°C) boiling-range solvent;
4. blowdown of the filter cake and depressurization with 200°F (92°C) nitrogen;
5. vacuum flash-drying to recover the solvent;
6. discharging of the filter cake; and
7. repressurization.

The filtered SRC solution goes to a vacuum-column preheater to boost the temperature to 600°F (315°C). Recirculated SRC is mixed with the filtrate and injected into the vacuum column, where the material is separated into liquid SRC which is drawn off at the bottom, solvent fractions and vapors in the overhead. The column overhead vapors are condensed and sent with the solvent-decanter stream to a light-solvent recovery column.

Liquid SRC is fed to two vibrating, water-cooled trays in the product cooler. The SRC solidifies into brittle sheets which shatter into small fragments upon vibration. The fragmented SRC is conveyed to storage.

Vapor from the high-pressure vent separator contains 60–80% hydrogen, plus hydrocarbon gases, hydrogen sulfide and carbon dioxide. The hydrogen sulfide and carbon dioxide are removed by scrubbing with a dilute solution of caustic soda. Scrubbed recycle gas is blended with pure hydrogen to provide an 85% hydrogen feed gas that is boosted by a hydrogen compressor to the inlet pressure of the slurry preheater.

Solvent from the vacuum column is fractionated into a bottoms having a boiling range of 450–800°F (232–427°C), which is used as the recycle process solvent, and an overhead fraction with a boiling range of 350–450°F (177–232°C) that is used in filter cake washing.

Coals tested in the SRC-I mode are given in Table 2-18. Modifications to the SRC process made since the plant finished construction have included numerous additions to the filtration step for removal of the ash from the SRC. Several centrifuge systems have been evaluated, with the feed material diluted with a wash solvent to reduce viscosity and with antisolvent added to enhance solids operation. The antisolvent and wash solvent are recovered in the overhead from the fractionation column. SRC containing less than 0.2% ash can be produced. Another ash separation technique used hydroclones to separate centrifugally the

Table 2-18. Coals Tested for SRC-I

Coal	Total Reactive Maceral Content (wt %)	Average Oxygen Content (wt %)	Volatile Matter (wt %)	H/C Atomic Ratio (average)	Iron Content of MF coal (wt %)	SRC Yield[a] MAF coal
Pittsburgh	93	7	39	0.80	1.0	61
Kentucky 9 and 14	95	8.5	38	0.85	2.7	59
Burning Star	95	9	38	0.83	1.2	58
Indiana V, High Vol. Bituminous, High S	92	11	37	0.85	2.4	55–60
Monterey, Ill. #6	94	10.5	43	0.86	1.5	56
Emery[b] High Vol. Bituminous Low S	79	9.5	39	0.83	0.3	45–55
Amax[b] Belle Ayr Wyoming Sub O.	93	19	50	0.86	0.2	44

[a]At 3H$_2$ consumption (MAF coal basis).
[b]Operating pressure 2400 psig; others at 1700 psig.

fractions without moving mechanical parts. Using antisolvents in the feed material, SRC with 1% ash can be produced. A critical solvent deashing unit has also been tested that produced SRC with less than 0.15% ash.

SRC-II Process Description

Because operations of the Wilsonville, AL, pilot plant experienced difficulty in separating the ash material, in the form of fine solids, from the solvent-refined coal, a modification was made to the process. By slurrying the pulverized coal with unfiltered coal solution, rather than with distilled solvent, as in SRC-I, the troublesome filters were bypassed. The ash was removed with the bottoms in a vacuum distillation step. The primary product of the SRC-II process was changed to a distillate with a reduced quantity of the solid fuel produced by SRC-I.

After a four-month period of plant modifications in 1977, the Fort Louis, WA, plant restarted production in the SRC-II mode as shown in Figure 2-9. Plant reliability was significantly better processing SRC-II than it had been in the SRC-I mode. Onstream time increased by one-third over the record achieved in SRC-I operations. The improved reliability was, in part, due to a reduction in the amount of equipment which must be operated in the SRC-II mode, compared to the more complex SRC-I operation. A new record of 61 days of continuous operation was set in late 1977.

MAJOR LIQUEFACTION PROCESSES 67

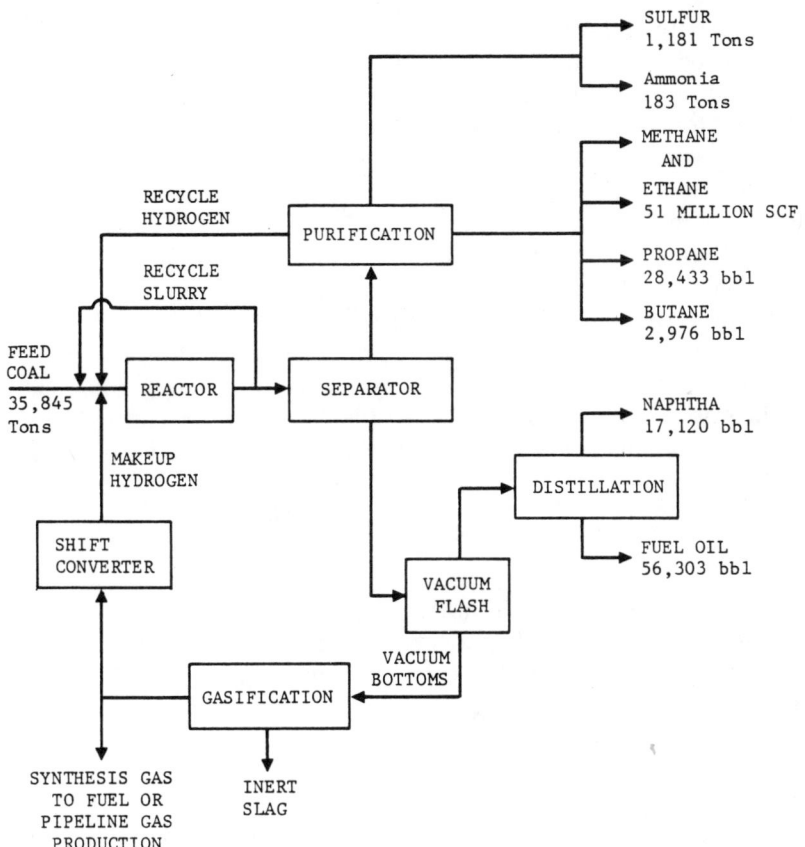

Figure 2-9. Simplified flowsheet for the SRC-II process.

Technical Evaluation

Historical Development

The original work on the solvent refining of coal dates back to the 1920s, when it was used in Germany for the production of material used to fabricate carbon electrodes for aluminum plants. The Spencer Chemical Company performed work on a modification of the Pott-Broche process. In 1962 OCR awarded a contract to Spencer Chemical to develop the solved-refined coal process. In 1965 the process was successfully demonstrated in a 50-lb/hr (23-kg/hr) continuous-flow process development unit. Spencer Chemical was acquired by Gulf Oil Corporation, and the SRC project was taken over by another Gulf subsidiary, the Pittsburg and Midway Coal Mining Company.

In 1966 a $28 million contract was awarded for construction and operation of a 50-ton/day (45-metric-ton/day) pilot plant that would be used for SRC process development, provide adequate samples of the SRC production and by-product for testing, and provide design data for scale-up to a commercial-size plant. Design of the plant began in 1972, and construction was completed in 1974 at Fort Lewis, WA. Operation at full design capacity was achieved a year later.

A smaller, 6-ton/day (5.4-metric-ton/day) pilot plant was constructed earlier at Wilsonville, AL, by Southern Company Services, under the sponsorship of The Southern Company, Edison Electric Institute and the Electric Power Research Insititute. The purpose of this plant was to obtain design data for the Fort Lewis plant and produce the solid product named SRC-I, which is used as a coal substitute in electric utility boilers.

Process Variation

Because of the change from the SRC-I to SRC-II mode of operation, a series of bench-scale tests was run to reevaluate the effect of process variables on the four coals used when operating in the SRC-II mode. The process variables studied were:

1. coal concentration in the feed slurry,
2. residence time,
3. dissolver temperature,
4. hydrogen feed rate, and
5. pressure.

Not all variables could be evaluated in full detail on each of the coals, but measurements were made on four bituminous coals:

Coal	Mine
Kentucky #9 and #14	Colonial
Illinois #6	River King
Pittsburgh	Ireland
Pittsburgh	Blacksbille #2

Coal concentration was found to have a substantial effect on distillate and residue yields, but no effect on the light hydrocarbon gas yields. As coal concentration increased, the process solvent and total distillate

MAJOR LIQUEFACTION PROCESSES 69

yields decreased linearly with coal concentration. Total distillation residue increased linearly with coal concentration.

The residence time of the coal slurry in the dissolver has a major effect on yields and can be varied over a wide range. A first-order kinetics model describes the disappearance of vacuum-residual material in the reactor and the production of distillate-range material. The results are consistent with a model in which the soluble (but not distillable) SRC is produced by a rapid reaction and disappear by a relatively slow reaction. The rates of production and disappearance of distillate materials are of similar magnitude. Light gas, process solvent and total distillate yields increase with increasing residence time, while SRC and total distillation residue yields decrease. Hydrogen consumption increases with increasing residence time.

One of the more important changes produced by the change from the SRC-I to the SRC-II was a significant increase in hydrogen consumption. While the SRC-I process only consumed about 2% of the moisture and ash-free weight of the coal, production of increased liquids in the slurry-recycle SRC-II mode will double or triple hydrogen consumption. Studies were performed to determine the minimum hydrogen feed rate that could be employed before product yields were seriously degraded or operability became a problem. As the hydrogen feed rate is decreased:

1. Hydrogen gas yield decreases.
2. Oil yield decreases.
3. Solid SRC yield increases.
4. Insoluble organic matter in the vacuum distillation residue increases.

Extrapolation of these results to larger-scale reactors is expected to hold quantitatively, but numerical results may show substantive differences because of differing reactor hydrodynamics that will affect the hydrogen feed rates. As hydrogen pressure is increased, the rate of conversion of asphaltene to oil goes up and oil yields improve, while SRC yield goes down.

Testing of different coals showed that satisfactory results were obtained from bituminous coals, but reactivity varied with rank of the coal and pyrite content. The pyrite serves as a catalyst to promote certain reactions in the dissolver. Coals deficient in iron can be treated with ferrous sulfate to substantially decrease SRC and insoluble organic matter yields and increase total distillate yield. Additional benefits of ferrous sulfate addition include increased hydrogenation, better nitrogen removal from the recycle solvent, and less sulfur and nitrogen in the

70 LIQUEFACTION

SRC. The yield of organic residue can be adjusted to the level necessary to produce hydrogen at the rate consumed by the process. The yield achieved with Kentucky Number 9 and 14 bituminous coals in both the SRC-I and SRC-II modes of operation is shown in Table 2-19 and

Table 2-19. Ultimate Analysis, Kentucky Coals No. 9 and 14, Colonial Mine[a]

	wt %
Carbon	71.17
Hydrogen	5.08
Nitrogen	1.50
Chlorine	0.04
Sulfur	3.37
Ash	11.41
Oxygen (by difference)	7.43
Total	100.00

[a]Source: Annual Technical Progress Report, 1977.

Table 2-20. Product Yields in SRC-I and SRC-II Modes (Kentucky Coals Nos. 9 and 14, Colonial Mine)[a]

	SRC-I Mode	SRC-II Mode
Yields		
H_2O	6.1	10.4
CO	0.4	0.4
CO_2	0.8	1.0
H_2S	1.9	2.3
NH_3	0.5	0.7
Light Gases		
C_1	3.7	5.7
C_2	2.9	4.1
C_3	2.3	3.7
C_4	1.6	2.6
Total C_1-C_4	10.5	16.1
Naphtha, C_5-193°C	10.6	10.0
Middle Distillate, 193-249°C	11.3	7.1
Heavy Distillate (Recycle Solvent), 249°C+	4.0	21.8
Total, C_5+	25.9	38.9
SRC	42.7	21.0
Insoluble Organic Matter	4.1	5.1
Ash	9.5	9.7
Total	102.4	105.6
H_2 Reacted	2.4	5.6
Moisture Ash Free Conversion (%)	95.5	94.4

[a]Source: Annual Technical Progress Report, 1977.

2-20 for runs with a 1-hour residence time, a dissolver temperature of 851°F (455°C) and a coal concentration of 30%.

Several tests were made on the suitability of a subbituminous coal from the Amax Coal Company's Belle Ayr Mine (Campbell County, WY). The principal objectives of these tests were:

1. to investigate the operability of the SRC-I process with a subbituminous coal;
2. to determine if recycle solvent balance could be maintained;
3. to investigate the extent of solids deposition in the dissolver; and
4. to develop process yield data under suitable operating conditions.

Due to the low reactivity of this coal, the reactor operating conditions for producing SRC-I varied significantly from those used for Kentucky No. 9. Residence time was increased to 1.3–1.5 hours, and pressure increased to 1900–2050 psi (133–144 kg/cm^2). The general yield pattern showed high, insoluble organic matter yields of 10–14%, and adequate recycle-solvent yield for a self-sustaining process. Solids accumulation in the dissolver eventually caused reactor plugging and terminated the runs. It was concluded that the Amax subbituminous coal was not an attractive candidate for the Fort Lewis pilot plant. Without a solids removal system or other technique for controlling solids accumulation, satisfactory operation could not be achieved.

Pilot-Plant Operations

Pilot-plant operations have encountered several technical problems with the SRC process. The dry coal is difficult to mix with the solvent and to pump, because of gel formation at low temperatures. The pilot plant has used two methods of mixing coal with the slurry at a temperature of 350°F (170°C). A third method of mixing at a temperature of 450°F (232°C) has been studied, together with an improved high-pressure pumping system. These improvements will be studied in pilot-plant tests. The formation of a gel as the coal dissolves increases the viscosity, thereby limiting heat transfer and increasing the pressure drop in the slurry preheater. Modifications to decrease the cost of commercial plant designs will be tested in the pilot plant.

The dissolver operation must control the temperature from the heat released by the exothermic reactions and maintain good contact between hydrogen and the slurry. The high temperature, high pressure and presence of solids in the slurry have caused problems with the dissolver effluent heat exchanger and erosion of the high-pressure letdown valves. The presence of solids creates additional problems in fractionation, particularly in pumping the bottoms from the vacuum tower. Pilot-plant operations have produced major improvements in the design of pumps for handling hot, concentrated slurries that have been successfully demonstrated.

Combustion Test Results

Although several liquefaction processes are being developed, the only pilot plant that has produced enough fuel for a meaningful combustion test in a utility boiler has been the Fort Lewis plant. An 18-day burn test of solid SRC-I was performed in 1977 at the Georgia Power Co. Mitchell Plant. The test results were successful although some modification of the equipment was required to handle the SRC. Windage losses of the ¼- to ⅛-in. material from open hopper cars were prevented by spraying the material with a commercial coating after loading. The ball-and-race mills used to pulverize the coal required three changes. Ambient air was substituted for hot air to keep the SRC from melting in the mill. The loading on the balls was reduced. The classifiers were modified to change the size range. A water-cooled specially designed nozzle was used in the boiler.

Boiler efficiency at full load was the same for SRC as it was with coal. Particulate loading at the electrostatic precipitator was 85–90% lower than when firing coal, but the fly ash was more difficult to collect because of the higher carbon content and low resistivity. Ash deposition within the boiler was low and easily removed. Because of the small capacity of the plant in comparison to the fuel requirements of a large boiler, many months of liquid SRC-II product were stockpiled for a commercial-scale burning test in 1978. A 3000-ton (2723-metric-ton) lot of SRC-II was produced for electric generation at the Plant Mitchell Station of the Georgia Power Company. A second lot of 4500 bbl of a blend of middle and heavy distillate was prepared for a large-scale combustion test by Consolidated Edison Company in New York. A 37-ton (33.6-metric-ton) lot of high ash vacuum bottoms was shipped to the Texaco pilot plant for tests as a gasification feedstock.

Alternative uses besides boiler fuel that are being considered for SRC-I include anode grade coke for aluminum smelting, metallurgical coke for steelmaking and other specialty carbon products. By adding a catalytic hydroprocessing step, distillate oils may be produced for chemicals and transportation fuels. Burning a slurry of oil and finely ground SRC-I as a boiler fuel is also under evaluation.

Commercial Plant Design

As part of the preliminary design of a 6000-ton/day (5442-metric-ton/day) demonstration plant to be located near Morgantown, WV, a conceptual design of a commercial-sized facility was also performed. The purpose of this design is to illustrate the technical problems and economics of using the SRC-II process at a commercial scale. The demonstration plant will consist of one train of equipment at the full commercial size. The commercial plant would consist of multiple processing trains with no increase in equipment size.

The plant design is based upon the use of Powhatan coal as given in Table 2-21. This West Virginia–Eastern Ohio bituminous coal produces the yields given in Table 2-22, similar to Illinois #6 and Kentucky coal. A block diagram of the plant design is given in Figure 2-20. The plant would have a daily product of the materials given in Table 2-23 and operate at a thermal efficiency of 72%.

The capital cost of construction is given in Table 2-24 with other charges determining total capital cost. The annual expenses are given in Table 2-25. Based on the above costs an economic analysis was performed that considered:

Inflation Rate	6%
Debit/Equity Ratio	25%/75%
Interest on Debit	9%
Investment Tax Credit	10%
Rate of Return on Equity	15%

Table 2-21. Coal Feedstock Composition (wt %, Moisture-Free Basis)[a]

Carbon	70.49
Hydrogen	4.88
Nitrogen	1.12
Oxygen	7.87
Sulfur	3.59
Pyrite	2.00
Organic	1.54
Sulfate	0.05
Chlorine	0.05
Ash	12.00
Contained Iron	1.75
Total	100.00

[a]Source: Phase Zero Conceptual Plant Design, 1979.

74 LIQUEFACTION

Table 2-22. Yield Data, Powhatan Coal Case[a,b]

Component	Yield (wt %)[c]
C_1	
C_2	4.4
C_3	4.1
C_4	2.2
C_5-350°F	7.9
350–600°F	20.1
600–900°F	8.0
900°F+	24.0
Undissolved Coal (IOM)[d]	4.62
H_2O	6.2
H_2S	2.45
NH_3	0.45
CO	0.08
CO_2	1.0
HCL	0.02
Ash	12.0
Residual S + Cl	1.08
Total	104.40

[a]Source: Phase Zero Conceptual Plant Design, 1979.
[b]Hydrogen consumption = 4.4 wt %.
[c]Moisture-free basis.
[d]Insoluble organic matter.

Table 2-23. Input, Output and Efficiency (Stream-Day)[a]

Inputs
Coal   1,388.875 ton/hr at 25.626 × 10^6 Btu/ton = 35,597 × 10^6 Btu/hr
Electricity: 10,429 kW at 9,500 Btu/kWh — 99 × 10^6 Btu/hr
Total Power: 36,696 × 10^6 Btu/hr

	Quantity/day	Heat Content (Btu/lb)	Output (10^6 Btu/hr)
Outputs			
Methane	51 × 10^6 scsd	23,120	1,992
Ethane/Propane	5,741,077 lb	22,008	5,265
Butane	878,561 lb	20,570	753
Naphtha	17,035 bbl	18,280	3,954
Fuel Oil	56,024 bbl	17,020	13,737
Total			25,701
By-Products			
Sulfur	1,175 tons		
Ammonia	182.6 tons		
Tar Acids	239 bbl		
Efficiency (Main Products Only) = 72%			

[a]Source: Phase Zero Conceptual Plant Design, 1979.

MAJOR LIQUEFACTION PROCESSES 75

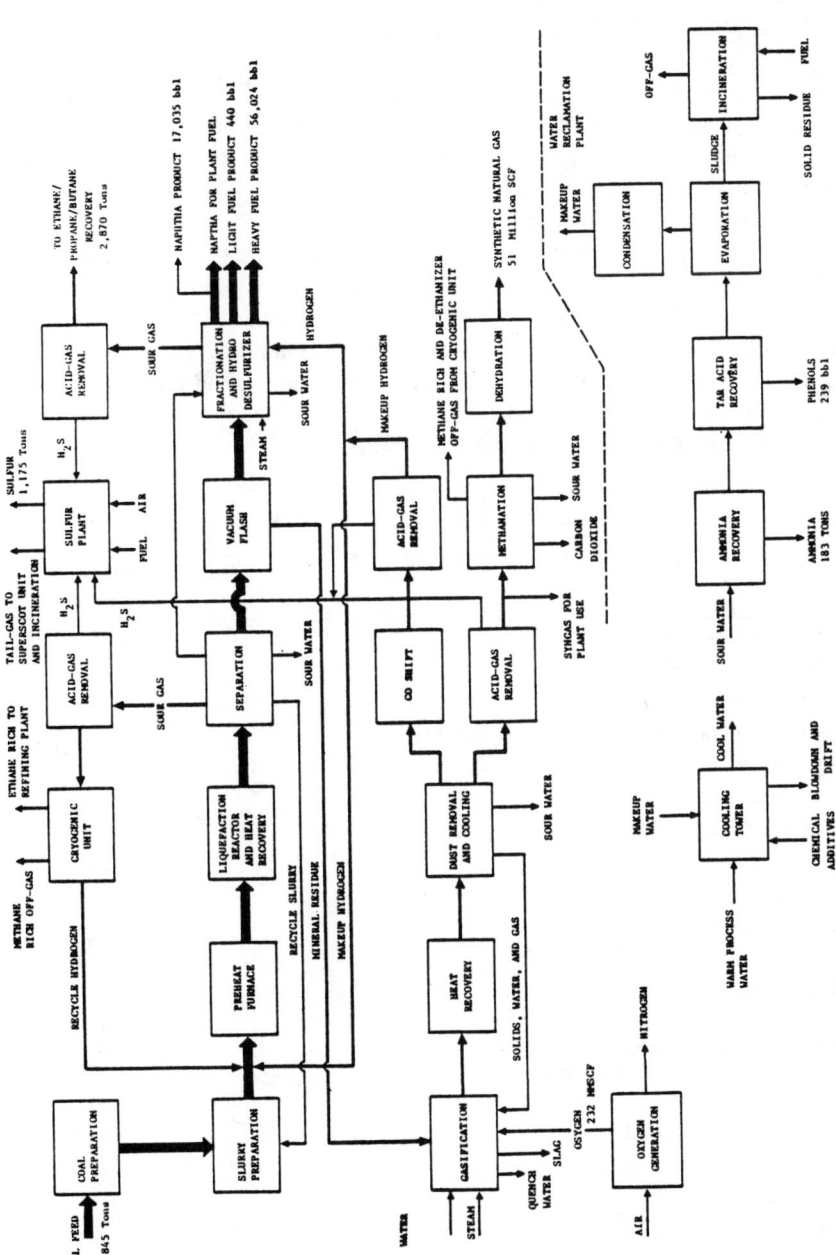

Figure 2-10. Simplified flow diagram for the proposed SRC-II commercial plant.

76 LIQUEFACTION

The analysis showed that the price of the products would have to be $3.761/10^6$ Btu ($14.92/10^6$ kg-cal) or $23–$25/bbl in fourth-quarter 1978 dollars.

Two auxiliary facilities that are not included in the SRC-II plant design are suggested by the developer to produce a finished product. A plant to upgrade the naphtha to produce high-octane blending stock for the production of gasoline would be accompanied by a second plant to reform the ethane-propane gas into ethylene.

The plant for upgrading naphtha would hydrotreat and reform 656,000 gal (2,483,000 liters) per stream-day into gasoline and is estimated to cost $44.3 million in November 1978 dollars. The ethylene plant would steam reform the gases into ethylene and propylene with butadine, aromatic distillate and high-purity hydrogen produced as by-products. A plant capable of producing 10^9 lb (454×10^6 kg) per year of ethylene is estimated to cost $209 million.

Table 2-24. Capital Costs (November 1978 Dollars)[a]

	Millions of Dollars
Unit 10: Primary Process Plants	615.2
Unit 20: Hydrogen Purification Plants	237.0
Unit 30: Refining and Gas Plants	140.3
Unit 40: Secondary Recovery and Oxygen Plants	262.3
Unit 50: Utility Systems and General Facilities	230.5
Unit 60: Coal and Ash Systems	74.7
Total Direct Capital Cost[b]	1,560.0
Catalysts and Chemicals	20.1
License Fees	13.2
Owner Management Costs	8.0
Land	7.8
Working Capital	
Raw Coal Inventory	26.9
Finished Products Inventory	10.6
Catalysts and Chemicals Inventory	.6
Spare Parts and Maintenance	24.8
Subtotal Working Capital	62.9
Total Indirect Capital Cost Elements	112.0
Total Capital Costs	1,672.0

[a]Source: Phase Zero Conceptual Plant Design, 1979.

[b]The capital cost has been estimated with a 20% contingency in accordance with DOE guidelines.

Assessment

The SRC-I and SRC-II pilot plants have acquired more operating experience in their size range, 6 and 50 ton/day (5.6 and 45 metric ton/day), than any other liquefaction technology being developed in the United States. Several years of operating experience has been accumulated by two contractors.

Scale-up of the process from 50 to 6000 ton/day (45 to 5400 metric ton/day) is in preliminary design. The size that would be achieved is about 20% of the size considered desirable for a commercial plant. The H-Coal and donor solvent pilot plants are of 250 ton/day (227 metric ton/day) capacity. A scale-up by the same size would result in a commercial plant.

In September 1979 the DOE announced that design would begin of two SRC demonstration scale plants. A 6000-ton/day (5443-metric-ton/day) SRC-I plant will be built at Newman, KY, 19 miles from the Ohio River. Four additional units can later be added to increase the capacity to a 30,000 ton/day (27,216 metric ton/day) commercial size. A similar approach will be followed for a SRC-II plant located near Morgantown, WV. The decision to construct the plants is scheduled for

Table 2-25. Annual Operating Cost (November 1978 Dollars)[a,b]

	Millions of Dollars
Coal	322.7
Direct Expense Elements	
Operating Labor	6.5
Operating Supplies	.6
Maintenance Labor	5.8
Maintenance Materials	31.8
Contract Maintenance	9.8
Catalyst and Chemicals	8.0
Electricity	3.0
Indirect Expense Elements	
Property Taxes and Insurance	23.4
Overhead	6.9
Total Annual Operating Cost Elements	418.5

[a]The annual operating cost has been estimated with a 20% contingency in accordance with DOE guidelines.
[b]Source: Phase Zero Conceptual Plant Design, 1979.

October 1980, with onsite construction beginning in early 1981. Plant operations will begin in late 1984.

BIBLIOGRAPHY

DeVaux, G. R., and B. Dutkiewicz. "H-Coal Commercialization," paper presented at the meeting of the American Institute of Chemical Engineers, November 1978.

"EDS Coal Liquefaction Process Development, Phase 111A. Annual Technical Progress Report, January 1-December 31, 1976," FE-2353-9, Exxon Research and Engineering Company, Baytown, TX (1977).

Epperly, W. R., and J. W. Taunton. "Development of the Exxon Donor Solvent Coal Liquefaction Process," paper presented at the 85th National Meeting of the American Institute of Chemical Engineers, Philadelphia, PA, June 7, 1978.

Epperly, W. R., and J. W. Taunton. "Exxon Donor Solvent Coal Liquefaction Process Development," paper presented at Coal Dilemma II, American Chemical Society, Colorado Springs, CO, February 12, 1979.

"H-Coal Commercial Plant. Part II Technical Proposal," Ashland Synthetic Fuels Inc. (1978).

Kunesh, J. G., et al. "Economics of the H-Coal Process," Hydrocarbon Research, Inc., Lawrenceville, NJ.

Lewis, H. E., et al. "Solvent Refined Coal Process. Quarterly Technical Progress Report, July-September, 1977," FE-2270-27, Catalytic Inc., Wilsonville, AL.

Lewis, H. E., et al. "Solvent Refined Coal Process. Operation of Solvent Refined Coal Pilot Plant, Quarterly Technical Progress Report for April-June 1978," Catalytic Inc., Wilsonville, AL.

"Phase II Laboratory Support for H-Coal Project," monthly progress report No. 14, FE-2547-14, Hydrocarbon Research, Inc., Lawrenceville, NJ (1978).

"Phase II Laboratory Support for H-Coal Project," monthly progress report No. 17, FE-2547-17, Hydrocarbon Research Inc., Lawrenceville, NJ (1978).

"Solvent Refined Coal Process. Annual Technical Progress Report, January-December 1977," Pittsburgh and Midway Coal Mining Co., Shawnee Mission, KS.

"SRC-II Demonstration Project, Phase Zero Conceptual Plant Design," Pittsburgh and Midway Coal Mining Company, Denver, CO (1979).

Swabb, L. E. "Liquid Fuels from Coal: From R&D to an Industry," *Science* 199:619-622 (1978).

CHAPTER 3
MINOR LIQUEFACTION PROCESSES

MOBIL GASOLINE SYNTHESIS

The Mobil process has been developed to maximize the production of light hydrocarbons that can be reformed into a gasoline motor fuel. The process catalytically dehydrates methanol to selectively produce naphtha, butane, propane, light fuel gases and water. After reforming, 85% by weight of the hydrocarbon product is gasoline, with 13% liquefied petroleum gas. The methanol can come from any source, but would be produced in practice by the gasification of coal and catalytic synthesis of raw methanol.

The Mobil gasoline synthesis system is schematically illustrated in Figure 3-1. Coal, oxygen and steam are gasified, purified and shifted to the optimum CO-to-H_2 ratio using commercially available technology. The synthesis gas is used to produce methanol, and the methane produced as a by-product can be utilized as a synthetic pipeline gas. The crude methanol is then treated by the Mobil process to produce water, gasoline, petroleum gas and a light gas that can be used to synthesize additional methane.

Because the methanol-to-gasoline conversion is separated from the methanol synthesis step, the Mobil research effort has only concentrated on process development unit and bench-scale work with the gasoline conversion reactor. No conversion of coal into methanol has been performed by Mobil. A 4-bbl/day fluid-bed reactor was constructed and has completed operations. Present commercial-scale methanol production plants are significantly smaller in size than would be required for synthetic gasoline production.

The advantages claimed in scaling up the Mobil process to a commercial size are that it:

80 LIQUEFACTION

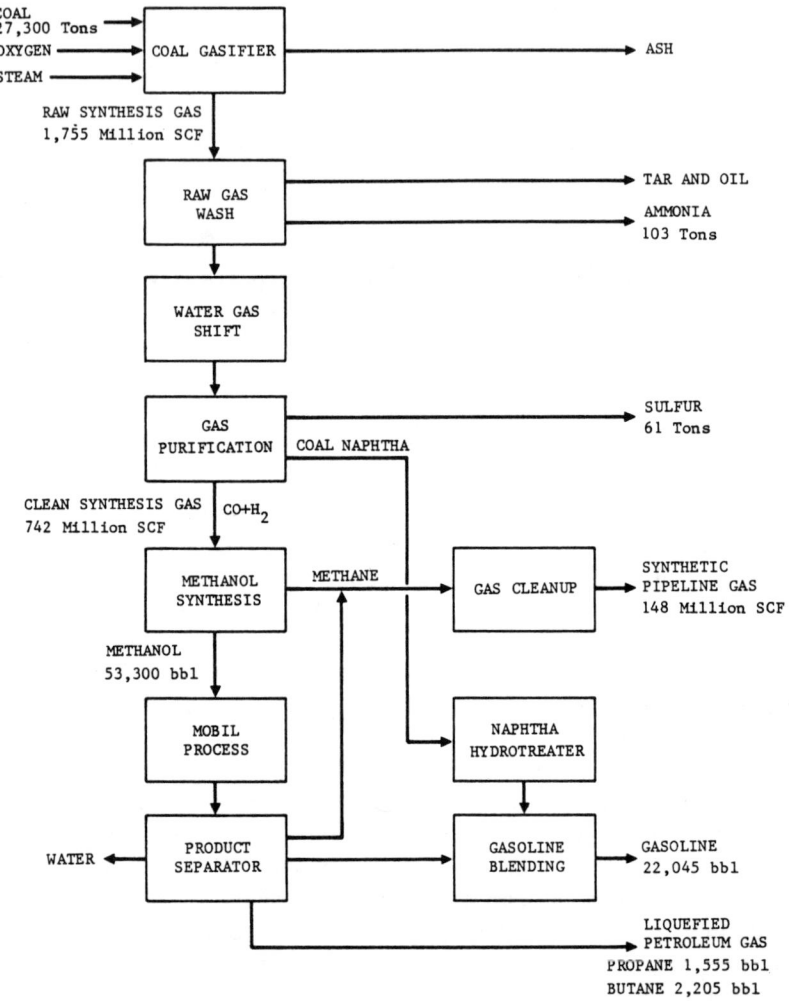

Figure 3-1. Mobil methanol-to-gasoline synthesis (source: Schreiner, 1978).

1. produces a high percentage of premium motor fuel;
2. has an acceptably high thermal efficiency;
3. has relatively few process steps;
4. can utilize existing technology for coal gasification and methanol synthesis;
5. utilizes existing petroleum refining technology for a large part of the plant;

6. can consequently be scaled up to commercial plant size with less risk than competing liquefaction technologies; and
7. will be able to utilize alternative processes or future improvements in coal gasification, purification or methanol synthesis.

Since the Mobil process begins with methanol feedstock, which could be directly used as a motor fuel or as a blending stock for gasoline, the advantages of converting it to gasoline should be explained to provide a justification for incurring the additional expenses. The biggest disadvantages are the low energy density of methanol and its high water solubility. Since a gallon of methanol has only one-half the energy content of gasoline, the volume of motor fuel to be shipped, stored and transported would be doubled. The mileage that a motorist would obtain from the fuel in his tank would be cut in half. Methanol, alone or in gasoline blends, tends to pick up water from the air. Other storage conditions can lead to a phase separation that separates the water-alcohol mixture from the fuel, resulting in engine shutdown. Additionally, methanol can prove corrosive to metals and cause swelling in elastomers. The expense of overcoming these problems (in terms of both energy and dollars) is greater than the 10¢/gal cost of methanol conversion into gasoline.

Process Description

The chemical reactions involved in the path from methanol to hydrocarbons begin with the conversion of methanol to dimethyl ether. Additional water removal produces olefins that are converted to paraffins, aromatics and cycloparaffins. A two-stage fixed-bed reactor system is used to limit temperature rise due to the exothermic reactions. The raw gasifier naphtha is hydrotreated under conditions that saturate the raw naphtha olefins with little aromatic-ring hydrogenation to reduce the octane rating as little as possible.

The principal problem to be overcome in the reactor design is removal of heat from the highly exothermic reaction. Conversion of methanol to hydrocarbons and water releases 740 Btu/lb. To minimize this problem, the reaction is divided into two steps. In the first step the methanol is reversibly dehydrated into dimethylether, releasing about 20% of the heat of reaction. In the second step the equilibrium mixture of methanol, dimethylether and water is converted to hydrocarbons and water. In the conversion reactor the Mobil catalyst converts the mixture of methanol and dimethylether first into light olefins and then by polymerization and rearrangement into higher-molecular-weight olefins, paraffins and aro-

82 LIQUEFACTION

matics. The reaction terminates at the upper end of the gasoline boiling range due to the unique nature of the catalyst. The Mobil process does not produce the heavy paraffins and alcohols that are by-products of the Fischer-Tropsch method of gasoline synthesis.

The technique is made possible by a zeolite catalyst, called the ZSM-5 class, that selectively converts methanol to the stoichiometric yields of hydrocarbons (44% by weight) and water (56%).

$$nCH_3OH \rightarrow (CH_2)_n + nH_2O$$

The action of the catalyst is highly selective in producing hydrocarbons in the gasoline boiling range (C_4–C_{10}). After reforming by conventional petroleum processing techniques, the gasoline produced has an unleaded octane number between 90 and 100 and is chemically similar to gasoline produced from petroleum. The catalyst is resistant to contamination during operations and can periodically be regenerated. After two weeks of operation enough coke accumulates on the catalyst to require regeneration. Combustion of the coke at a controlled temperature restores the activity. Since the catalyst is tolerant of water, crude methanol can be used without any purification.

The equipment used in the process (Figure 3-2) is similar to a two-stage petroleum hydrotreater. Crude methanol is first passed through

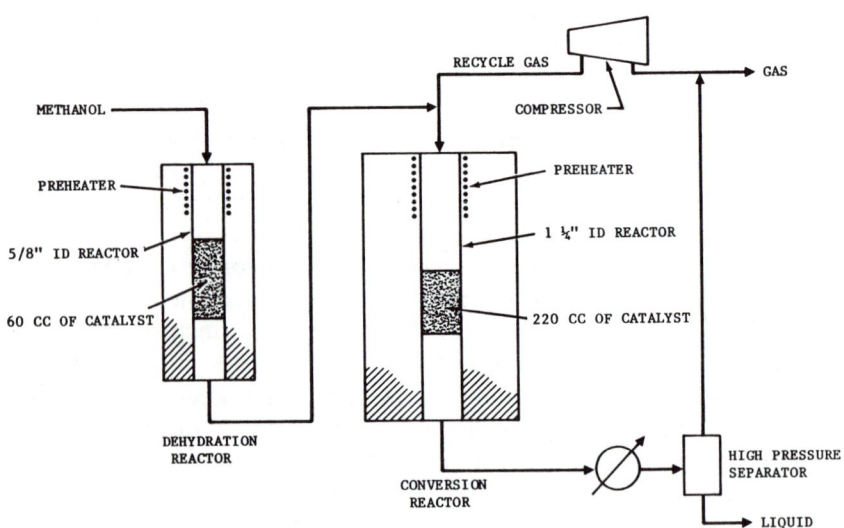

Figure 3-2.   Schematic of fixed-bed pilot plant.

a dehydration reactor and then into the conversion reactor. Recycle gas provides a heat sink to pick up the heat of reaction and limit the temperature rise. The gasoline conversion reactor inlet temperature is typically 680°F and with a 9-to-1 recycle ratio, the adiabatic temperature rise is about 100°F. The reactor effluent is condensed, water and liquid hydrocarbon phases are separated, and the gas is recycled. Liquid and gaseous hydrocarbons are processed through a conventional gas plant to make additional gasoline. Propene, butenes and isobutane are separated and alkylated using conventional pertoleum technology.

The gasoline produced by the process is of very high quality, consisting of highly branched paraffins and olefins, naphthenes, and aromatics. The gasoline does not contain sulfur or nitrogen since there is none in the methanol feed. A gasoline yield of 85% by weight is achieved with 13.6% by weight of liquefied petroleum gas produced.

The thermal efficiency of the Mobil process using methanol produced from coal is slightly less than the efficiency that can be achieved by other coal liquefaction technologies. The thermal efficiency of the Mobil process for converting methanol into gasoline is 88%. However, the overall process efficiency also depends on the coal gasification process and the by-products that are produced. Three options for plant design have been compared. The first option is the production of synthetic pipeline gas plus by-products such as naphtha and tar by the Lurgi process. The second is to produce methanol plus synthetic pipeline gas and by-products. The final option is to add on the Mobil conversion process and optimize the system for maximum financial return by adding methane and petroleum gas to the list of final products. This third option produces an output consisting of 46% substitute pipeline gas, 31% gasoline, 5% petroleum gas and 18% by-products. The reduction in thermal efficiency between the first and third options is only 3.3%.

Preliminary Design of a Commercial Plant

The source of information for the Mobil process on a commercial scale comes from a detailed design study prepared by Mobil with the assistance of the American Lurgi Corporation, for the U.S. Department of Energy (DOE). The purpose of the study was to provide a technical and economic comparison between the Mobil methanol to gasoline technology and the Fischer-Tropsch (F-T) technology. The F-T process is commercially available and is used on a commercial scale by Sasol in South Africa. Two conceptual plant complexes processing 25,000 ton/day of Western coal were designed to produce a mixture of synthetic

84 LIQUEFACTION

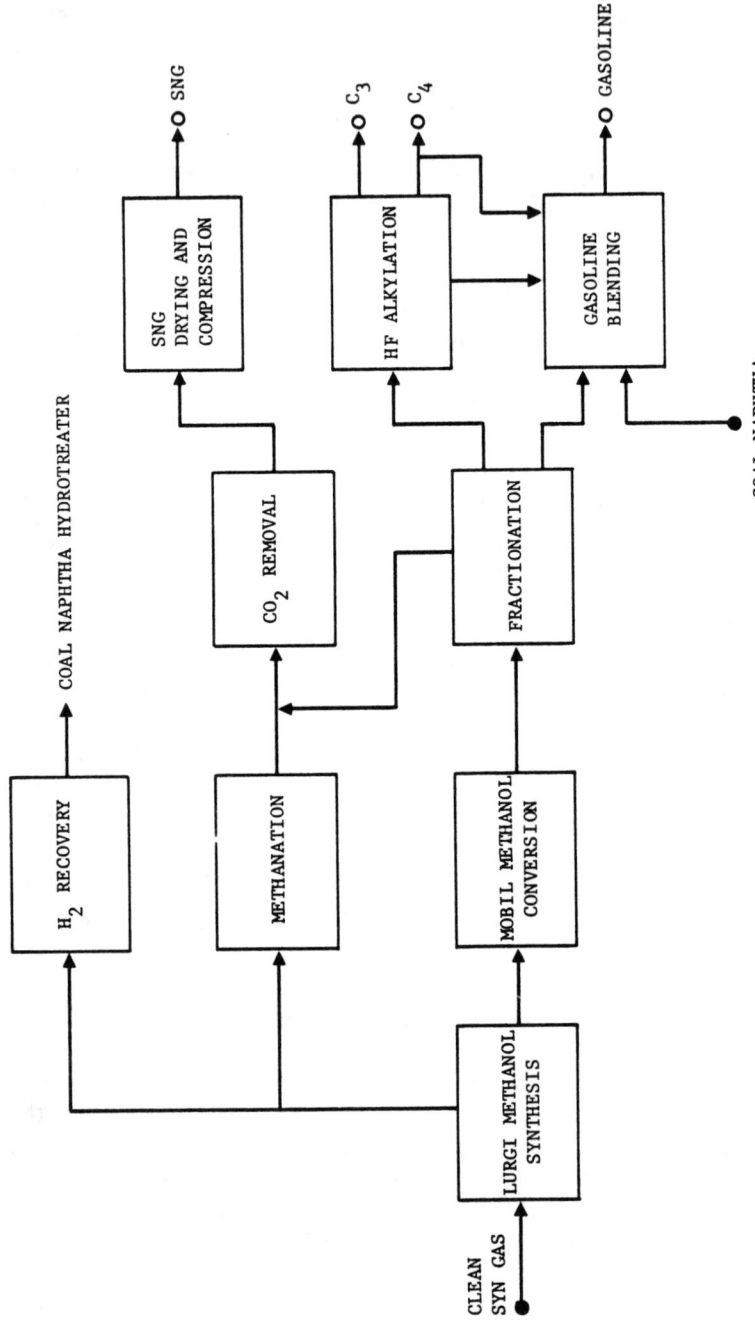

Figure 3-3. Mobil methanol-to-gasoline process (source: Schreiner, 1978).

pipeline gas, liquefield petroleum gas, gasoline and by-products. The Lurgi dry-ash process was first used to gasify subbituminous strip-mined coal. After gas cleanup the synthesis gas was fed into the process steps of the Mobil (Figure 3-3) and F-T processes. Estimates were made of the cost of equipment, processing steps, the quantities and quality of the products produced, and the required selling prices of the products.

Mobil Commercial Plant Design

Coal is crushed and sized for gasification with the fines fraction rejected. The feed coal is charged into the pressurized dry-ash gasifier through lock hoppers. As the coal passes through the gasifier, a countercurrent flow of gas dries the coal, devolatilizes and gasifies it. Oxygen and steam are injected into the bottom of the gasifier through a rotating grate; ash is removed through a lock hopper. An excess quantity of steam is required to limit the gasifier temperature below the ash liquefaction point. The hot raw gas leaving the gasifier is cooled and scrubbed with recycled gas liquor in a wash cooler. The crude gas is saturated with steam and the dust and heavy tars are removed. Further cooling takes place in a waste heat boiler where 100-psig steam is produced. Part of the raw gas is sent to a shift converter where the H_2/CO ratio of the raw synthesis gas is increased to meet the requirements for methanol conversion. The converter catalyst is periodically regenerated.

Gas from the shift converters and gasifiers is cooled in several heat exchangers and steam generators before entry into the low-temperature Lurgi Rectisol gas purification process. For use with low-sulfur coal, the Lurgi Rectisol process combined with the Stretford sulfur recovery method is the most economical system. The Lurgi Rectisol process uses methanol refrigerated to 32° and −45°F to remove naphtha, HCN, water, sulfur compounds and carbon dioxide. Naphtha, sulfur, ammonia and synthesis gas are recovered from the methanol. Tar, oil and phenol are recovered from the scrubbing liquors and utilized as boiler fuel.

The Lurgi low-pressure methanol synthesis process is used to produce methanol from the clean synthesis gas. The heat liberated in the reaction is removed from the catalyst bed with a steam generator. The methanol that is produced goes to the reactor for conversion into gasoline by the Mobil process. The unreacted gas is processed to recover unreacted hydrogen, remove CO and CO_2, and synthesize methane for use as a synthetic pipeline gas. The Union Carbide pressure swing process recovers 98% pure hydrogen for use in hydrotreating the gasifier naphtha. Methanation occurs in the Lurgi methanation process cata-

86 LIQUEFACTION

lyst reactor. The Union Carbide amine guard process uses monoethanolamine to absorb CO_2 and lower its concentration to less than 0.5%. A triethylene glycol commercial process is used to dry the gas before introduction into the pipeline system.

Crude methanol from the synthesis unit is pumped to the reaction pressure of 315 psig and vaporized in a heat exchanger. In the first stage of a two-stage reactor, the methanol is catalytically converted into an equilibrium mixture of methanol, dimethylether and water. The equilibrium mixture flows out of the first stage and into the second stage for conversion into gasoline. It is mixed with a recycle gas stream to control the reactor inlet temperature to 625°F. The methanol is converted into water, light petroleum gases (C_1-C_4) and a gasoline consisting of:

- highly branched paraffins, 51%;
- highly branched olefins, 13%;
- naphthenes, 8%; and
- aromatics, 28%.

No hydrocarbons larger than C_{10} and no oxygenates are produced. Small amounts of CO, CO_2 and coke, and trace amounts of formic acid and acetone are formed as by-products. After 14 days onstream, enough coke accumulates on the catalyst to require regeneration by combustion. Additional gasoline is formed by hydrotreating the raw gasifier naphtha and converting isobutane and butene/propene into commercial liquefied petroleum gases and gasoline by alkylation.

Supporting plant process and facilities include:

- 98% purity oxygen plant,
- high-pressure steam coal-fired boiler and superheater,
- lime slurry scrubber and electrostatic gas precipitator,
- instrument and plant compressed air,
- coal crushing, screening and storage facility,
- ash handling and disposal,
- fresh water and boiler feed water treatment plants,
- wastewater treatment plant,
- electric power generating unit,
- gas flare and blowdown holding pond, and
- product storage facilities.

The properties of the Wyoming coal on which the plant is based are given in Table 3-1 and the plant mass balance is given in Table 3-2.

The process information was based on either published or licensed data. All technologies used in the plant are commercially available.

Table 3-1. Wyoming Subbituminous Coal Properties

	As Received	Moisture and Ash Free
Proximate Analysis (wt %)		
Moisture	28.0	
Ash	5.1	
Fixed Carbon	33.8	50.5
Volatile Matter	33.1	49.5
Total	100.0	100.0
Ultimate Analysis (wt %)		
C		74.45
H		5.10
O		19.25
N		0.75
S		0.45
Total		100.0
Calorific Value (Btu/lb)		
High Heating Value	8,509	12,720
Low Heating Value	7,893	12,236
Calorific Value (J/kg)		
High Heating Value	19,768	29,510
Low Heating Value	18,312	28,388

Much of the coal gasification information was derived from proposed Western synthetic pipeline gas plant studies. Combining the processes to obtain a combination of output products was based on the experience of the study participants. The coal gasification, synthesis gas purification, methanol synthesis and methanation are commercially available processes of Lurgi Kohle and Mineraloeltechnik. The data for the F-T technology are derived from references in the published literature. Sasol did not participate in the study, so the exact Sasol commercial technology cannot be determined. Investment estimates were derived from Mobil in-house data.

Comparison of Mobil and Fischer-Tropsch Processes

Six process configurations were analyzed, including the use of Mobil fixed and fluid-bed reactors, Lurgi and Mobil methanol synthesis processes and a Mobil direct gasoline synthesis process that does not use the methanol step. Reforming of light hydrocarbon gases to produce only liquids was analyzed. The work performed included determination of:

88 LIQUEFACTION

- material balance,
- utility balance,
- plant thermal efficiency,
- block flow diagrams,
- simplified process flow diagrams, and
- equipment lists.

Two base cases examined in detail used the Mobil methanol conversion with a fixed-bed reactor and the F-T process with the Sasol-II type fluid-bed reactor. The analysis considered nine processing steps common to both technologies required for coal gasification, cleanup and synthesis gas production. Nine additional steps are required for the Mobil process to produce marketable products. The F-T process requires 18 processing steps. Both processes require 20 types of supporting facilities and 14 infrastructure units to carry out commercial operations.

Both processes produce several forms of fuel as gasoline, synthetic pipeline gas, and liquefied petroleum gases. The F-T process additionally produces heavier oils and alcohols that have fuel or feedstock value.

Table 3-2. Mobil Methanol-to-Gasoline Commercial Plant Mass Balance (Stream-Day Basis)[a]

Coal Handling Facility Input (coal as mined)		27,334	tons
Process Input			
Moisture- and Ash-Free Coal (per Table 3-1)		18,286	tons
Gasifier Input Coal		15,284	tons
Boiler Input Coal		2,179	tons
Raw Water Input		37,725	tons
Process Output			
Synthetic Pipeline Gas		148×10^6	scf
Gasoline		22,045	bbl
Butane		2,205	bbl
Propane		1,555	bbl
Sulfur		61	tons
Anhydrous Ammonia		103	tons
Excess Electric Power		5.31	MW(e)
Coal Fines, Moisture- and Ash-Free		823	tons
Thermal Conversion Factors			
Coal HHV	12,720 Btu/lb	7,060	cal/kg
Synthetic Gas	980 Btu/scf	36.75	J/normal m³
Gasoline	5.1×10^6 Btu/bbl	5.38×10^6	kJ/bbl
Butane	4.2×10^6 Btu/bbl	4.43×10^6	kJ/bbl
Propane	3.8×10^6 Btu/bbl	4.0×10^6	kJ/bbl
Thermal Efficiency			62%

[a]Source: Schreiner, 1978.

The yield of gasoline is about 60% greater in the Mobil process and has less of the thermal energy in synthetic pipeline gas. The Mobil process efficiency of 62% is slightly better than the F-T process efficiency of 58%.

The Mobil process is simpler, produces better-quality gasoline, and selectively directs the products produced to obtain a 60% higher yield of premium transportation fuel. The proportion of products produced is chosen to maximize plant efficiency and minimize cost. For both processes this results in a significant fraction of output in the form of synthetic pipeline gas with a price that was not competitive with natural gas in the U.S. prior to 1980. The Mobil process could be altered to reform the methane and produce additional gasoline, but at a sacrifice of 15 percent in thermal efficiency and an increase of 26 percent in gasoline cost.

Economic Analysis

An extensive economic analysis was performed using both utility-type and equity financing. The prices received for by-products were analyzed on a thermal value basis and on the basis of "Gas Cost Guidelines" developed to standardize the analyses of coal gasification projects. Tables 3-3 and 3-4 list the assumptions and investment costs. Tables

Table 3-3. Investment Cost Assumptions[a]

Pricing based on October 1977 costs
Direct construction labor average usage rate—$11.38/man-hour
Field indirect costs are 110% direct field labor costs
Freight on equipment and materials—6%
Home office engineering and design costs are 11.5% field construction costs.
Construction labor camp is 6% of contractor costs
Double pay overtime premium paid for 14 hours/week
Other project costs include: sales tax, contractor fee, spare parts, project management costs, builder's insurance, catalyst and chemicals, and process royalties
Items excluded: land costs, inflation, special construction foundation conditions, utility tie-ins, and royalties for Mobil and Fischer-Tropsch technologies
Expense capital includes: sales tax, catalyst and chemicals, and environmental studies
Working capital includes: 30-day coal inventory, one-half storage capacity product inventory, materials and supplies, net receivables, and cash-on-hand

[a]Source: Schreiner, 1978.

90 LIQUEFACTION

3-5 and 3-6 list the assumptions and operating costs. Tables 3-7 and 3-8 list the assumptions and unit operating costs. Mobil includes in its report a caution against making an economic comparison between the results of this report and results found in the literature. The cost data

Table 3-4. Investment Costs ($ million)[a]

	Mobil Methanol	Fischer-Tropsch
Process Units		
Gasification	430	430
Gasoline Production	132	180
Synthetic Gas Production	26	26
Offsite Units		
Oxygen Facilities	110	110
Steam Facilities	126	148
Water Facilities	68	76
Catalyst Preparation		28
Other	146	142
Infrastructure	46	46
Total Field Construction	1,084	1,186
Engineering and Design	139	152
Other Project Costs	248	270
Estimating Allowance at 15%	220	242
Process Royalties	17	13
Total Depreciable Capital	1,708	1,863
Expense Capital	24	24
Working Capital	58	60

[a]Source: Schreiner, 1978.

Table 3-5. Operating Cost Assumptions[a]

Wyoming strip-mined subbituminous coal—$7/ton
Operating labor is 50% higher for F-T process due to higher process complexity
Hourly wage rate—$7.85/hr
Burden factor—20%
Supervision overhead—20%
Annual wage rate including burden—$26,450/yr
Maintenance cost is 3.5% of construction cost (labor—60%, materials—40%)
Water cost—$0.45/1000 gal
Purchased electric power—1.65¢/kWh
Supplies cost is 30% of operating labor and maintenance
Administration and general overhead is 60% of total labor cost
Local taxes and insurance is 2.7% of total capital cost
By-product credits: sulfur, $25/ton; ammonia, $140/ton; electric power, 0.45¢/
    kWh; coal fines, $5.25/ton

[a]Source: Schreiner, 1978.

for other processes are often rough scoping estimates useful for planning and technical sales. The plant engineering on which they are based is conceptual and contains optimistic cost estimates made in accordance with unrealistic guidelines.

The cost comparisons indicated that the Mobil process could produce gasoline in October 1977 dollars for $0.60–$1.00/gal refinery

Table 3-6. Operating Costs ($ million per year)[a]

	Mobil Methaonl	Fischer-Tropsch
Labor and Supervision		
Operating	6.9	10.2
Maintenance	29.5	33.2
Water	1.4	1.4
Supplies	21.4	24.7
Catalyst and Chemicals	5.0	5.9
Administration and Overhead	21.9	26.1
Local Taxes & Insurance	45.6	50.0
Total operating costs	131.7	151.5
Coal[b]	64.3	64.3
By-product Credits	(8.2)	(6.0)

[a]Source: Schreiner, 1978.
[b]Credit accepted for coal fines sold as boiler fuel.

Table 3-7. Economic Analysis Assumptions[a]

	Equity Financing	Utility Financing
Plant Operating Life	20 years	20 years
Depreciable Life	13 years	20 years
Construction		
Studies, Design and Engineering	2 years	
Erection	4 years	
Investment Credit	7%	Not used
Depreciation Method	Sum of the year's digits	Straight line
State and Federal Income Taxes	51.9%	
Startup Penalty	One half production in first operating year	20% annual costs
Debt/Equity Ratio		75%/25%
Unit Cost Basis		9% interest in debt 15% return on capital

[a]Source: Schreiner, 1978.

price, while the F-T process would cost $0.70–1.35/gal on comparable bases. The Mobil process cost was about 25% lower. Plant costs were high, about $1.8 billion for a Mobil process plant producing 22,000 bbl/day of gasoline, or 13,600 bbl/day of gasoline by the F-T process and a total fuel oil equivalent output of all forms of fuels of 45,000 bbl/day for both processes.

Inexpensive Wyoming strip-mined coal, at $7/ton, was used for the analyses. High capital costs predominated in the analyses, accounting for 60–65% of the unit cost. Annual operating costs were 20–30% of unit costs; coal costs only amounted to 10% of unit cost.

The synthetic pipeline gas represents half the plant output and has been priced at the value of synthetic gas, estimated from a dedicated plant using the Lurgi process to be $6.17/$10^6$ Btu. Because the estimates are made for a time frame of 10–15 years in the future, large uncertainties exist in extrapolating historical refinery price differences for different products, different product quality differences and the overall product market structure. Then, depending on the pricing and financing assumptions, the refinery selling price for the Mobil methanol conversion technology is estimated to range between $0.60 and $1.00/gal. Operation of the plant in a gasoline-only mode, with no synthetic pipeline gas production, increases the gasoline cost by $0.15–0.25/gal. The Mobil methanol conversion technology has a cost advantage of $0.06–$0.40/gal over the commercial F-T technology. For comparison, the

Table 3-8. Unit Costs[a]

	Mobil Methanol	Fischer-Tropsch
Equity Base @ 12% Discounted Cash Flow		
Capital charges	4.59 per 10^6 Btu	5.07 per 10^6 Btu
Startup penalty	0.35	0.38
Coal cost	0.70	0.72
Operating expenses	1.44	1.68
By-product credit	(0.09)	(0.07)
Total cost	6.99	7.78
Utility Base		
Capital charges	2.98	3.28
Startup penalty	0.05	0.06
Coal cost	0.70	0.72
Operating expenses	1.44	1.68
By-product credit	(0.09)	(0.07)
Total cost	5.08	5.67

[a]Source: Schreiner, 1978.

refinery price of gasoline in the United States in May of 1979 was $0.51 to $0.62 per gallon.

Project Assessment

The Mobil methanol conversion process produces a premium motor fuel which satisfies the largest segment of the American liquid fuel market. Because of the great demand a high price could be supported. The economics of the process is dependent on the conversion cost of coal to methanol. These costs have been estimated for operations at a scale considerably larger than any existing commercial practice. The most efficient and most economical process design also produces significant quantities of synthetic pipeline gas and liquefied petroleum gas. There is no American market for the synthetic pipeline gas at the prices used in the economic analyses and only a small market for the petroleum gases. A 100-bbl/day pilot plant for methanol conversion is in the planning stage. Construction and operational experience will not be available for another 5 years. A commitment to construct a commercial plant would involve accepting the risks of an extremely large scale-up over demonstrated technology and operations. The W. R. Grace Co. has been awarded a contract to perform a conceptual design of a plant located in Kentucky, which would utilize the Mobil process. The plant would cost $3 billion and convert 29,000 ton/day (26,310 metric ton/day) of coal into methanol, and then into 50,000 bbl/day of gasoline.

FISCHER-TROPSCH SYNTHESIS

The F-T method of indirect hydrocarbon synthesis is one of the oldest processes. The development work was performed in the 1920s, with commercial plants operating before World War II. The German petroleum shortage during the war resulted in its greatest utilization, when nine plants using F-T technology accounted for 12% of the total peak German production of 100,000 bbl/day. By 1962 all of the plants had ceased production because of the availability and low cost of petroleum. The Union of South Africa was faced with the necessity of having an internal petroleum source and chose the F-T process. Production from the Sasol plant began in 1955. A production of 6000 bbl/day from 6600 ton/day (5991 metric ton/day) of coal satisfied about 10% of the country's motor fuel needs. A second plant, processing 38,000 ton/day (34,496 metric ton/day) will begin operation in 1980 and increase pro-

94 LIQUEFACTION

duction to about half of the national consumption. A third plant is being planned using the same process design as the second.

The F-T process gasifies the coal into a synthesis gas of carbon monoxide and hydrogen, cleans the gas of impurities and feeds it into a catalytic reactor, where it combines into long-chain aliphatic hydrocarbons including gasoline, alcohol, waxes, oil and gas. The products are separated and upgraded, with fuel gas being produced as a byproduct. Additional materials obtained during gasification include ammonia, phenols, sulfur, naphtha, coal tar and oils. The F-T reaction uses powdered iron catalysts in an Arge fixed bed or, in the newer units, a Synthol entrained-bed reactor.

Advantages of the process include the avoidance of many problems associated with coal handling in the early stages of direct liquefaction by converting it into synthesis gas. Other types of organic matter that could be gasified are also potential feedstock material. A broad range of products are produced from the reactor, some of which do not have ready markets when produced in large quantities. The yield of gasoline and motor fuels per ton of coal consumed is relatively low, and the gasification step and high heat rejection during catalytic synthesis results in a low thermal efficiency.

Fischer-Tropsch Process Description

This description is taken from the Mobil study of comparative methanol-to-gasoline and Fischer-Tropsch technologies.

The coal gasification section of the Fischer-Tropsch plant is identical to that of the Mobil methanol conversion process described above, consisting of:

- coal storage and handling,
- Lurgi gasifiers,
- raw gas shift and cooling,
- gas purification,
- gas liquor separation, and
- sulfur, phenol and ammonia recovery.

A heavy hydrocarbon recovery unit is used to remove heavy hydrocarbons from the Fischer-Tropsch synthesis purge gas to meet methanation feedstock specifications. A low-temperature heptane wash was selected because of the CO_2 content of the gas.

The Union Carbide pressure-swing process is used to absorb all of

MINOR LIQUEFACTION PROCESSES 95

the nonhydrogen components of the feed gas to produce 98% pure hydrogen for use in the plant hydrogenation/hydrotreating units. The gases desorbed from the process are recompressed and returned to the synthetic natural gas (SNG) upgrading train as feed to the processing unit. The Lurgi methanation process is used to produce synthetic pipeline gas with a CO content of less than 0.1%. The gas that is produced is reduced in CO_2 content to less than 0.5% by scrubbing with monoethanol amine and reduced in water content by scrubbing with ethylene glycols.

The design of the hydrocarbon synthesis unit illustrated in Figure 3-4 is based on the Sasol technology, as reported in the literature. A full range of products can be produced, including methanol, liquid hydrocarbons and waxes. The product includes paraffins, monoolefins, aromatics, alcohols, aldehydes, ketones and fatty acids, with carbon numbers from 1 to more than 35. Small amounts of diolefins and esters can also be produced. The structure is predominantly a single-methyl branched structure when it occurs. Yield, selectivity and composition are highly dependent on the catalyst, reaction conditions and reactor type.

Commercial catalysts include cobalt (fixed-bed) and iron (fixed- and fluid-bed). Both are promoted for improved activity and selectivity. Operating conditions range from 392 to 617°F (200 to 325°C) and from 1 to 25 atm pressure, depending on products desired, catalyst, and reactor design. Because the reactions that occur are highly exothermic, the principal problem in designing the reactor is heat removal. Fixed-bed reactors have a heat exchanger, with the catalyst cooled by boiling water or circulating oil. Fluid-bed reactors have internal-tube bundles for removal of reaction heat. The fluid-bed design gives the following products:

- C_1 to C_4: 43%
- C_5+: 8%
- light oil: 35%
- heavy oil: 7%
- alcohols: 6%
- acids: 1%

The design estimates that 85% conversion of CO and H_2 can be obtained at operating conditions of:

- vapor feed temperature: 320°F (160°C)
- catalyst and vapor outlet temperature: 644°F (340°C)
- vapor outlet pressure: 300 psia (21 kg/cm^2)
- catalyst life: 50 days

96 LIQUEFACTION

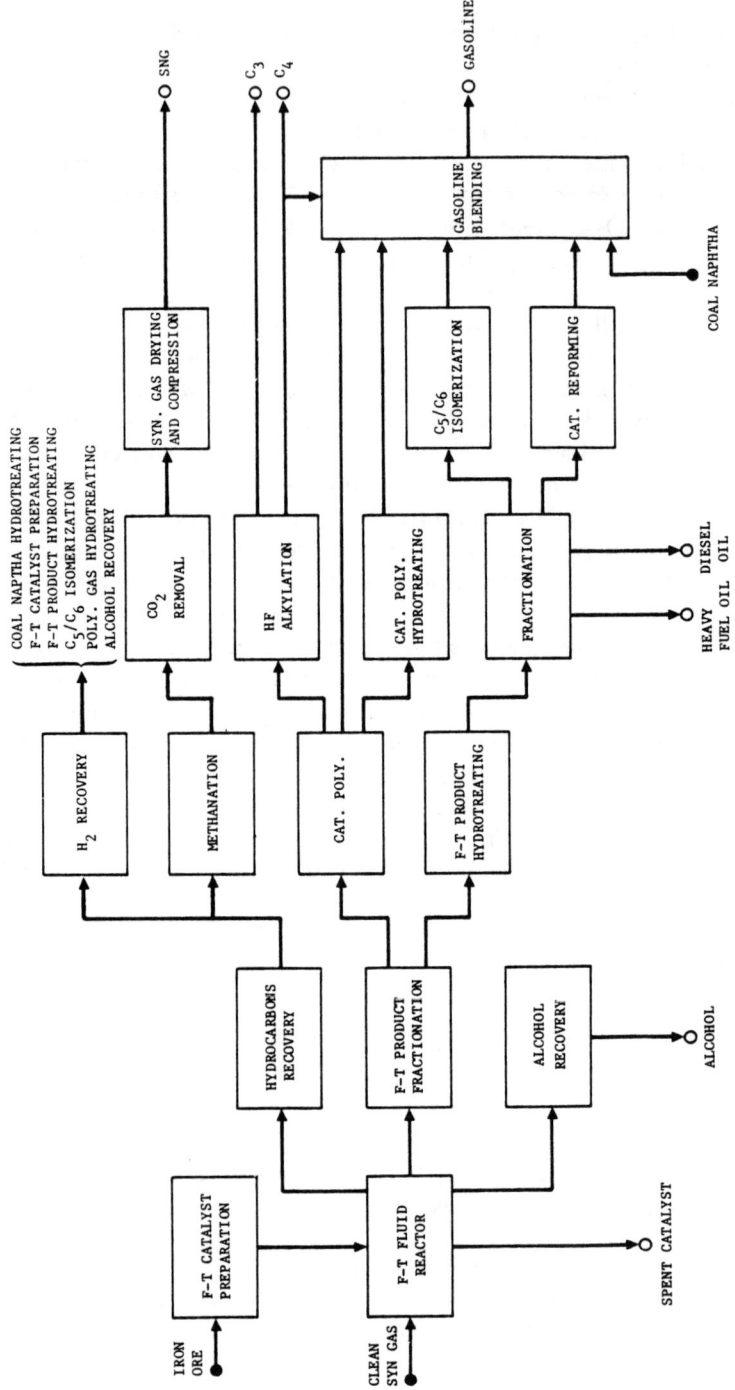

Figure 3-4. Fischer-Tropsch synthesis (source: Schreiner, 1978).

A product-fractionating unit is used to separate the output stream into light gases, feedstocks for the polymerization and hydrogenation units, and an 850°F+ (440°C+) residual for boiler fuel. The naphtha is hydrotreated to yield a clean, saturated, stabilized gasoline pool-blending feedstock. A hydrotreater is used to saturate the olefins and to destroy the remaining alcohols and acids. A product-fractionating unit is used to separate the hydrotreated product into a pentane/hexane stream for isomerization, a C_7-380°F (C_7-193°C) naphtha for reforming, a diesel oil product, and a heavy fuel oil product. The C_7-380°F naphtha is reformed to increase the antiknock quality by platinum catalysts that stimulate aromatic formation. Pentanes and hexanes are catalytically isomerized to increase the octane number. Propene/butene mixtures from the F-T reaction are polymerized into higher-molecular-weight compounds suitable for gasoline blending. Isobutane and unpolymerized C_3 and C_4 olefins are catalytically alkylated to increase the gasoline yield. The heavy gasoline fraction is separated at the C_9-C_{10} boiling range and hydrogenated to saturate the olefins. Light gasoline and isobutane are separated from the hydrocarbon-rich offgases. Alcohols are separated from the water to produce a marketable alcohol mixture and to recover methanol for makeup to the Rectisol unit. Hydrogenation of the aldehydes and ketones improves the marketable value of the alcohol mixture. A cryogenic process is then used to produce a hydrogen makeup stream of 90% purity.

The product yield per stream day for the Fischer-Tropsch plant is given in Table 3-9.

The diesel oil is a premium fuel suitable for engine service requiring frequent speed and load changes. Because the heavy fuel oil has nonexistent sulfur and metals content, it could be used as a premium gas turbine fuel. The process has been adjusted so that the gasoline produced is 87 octane, and slightly lower in gravity and containing fewer aromatics than typical, present-day gasolines. A high olefin content of 20 percent may produce problems that will require extensive testing to resolve.

Assessment

Fischer-Tropsch synthesis has been technically developed to a high degree and has been in commercial usage for more than 25 years. Additional improvements in the synthesis portion of the process will be difficult. The present Sasol plant uses Lurgi dry-ash gasifiers. Economic im-

98 LIQUEFACTION

provements could be achieved by using an improved gasifier. Analysis shows that more than half of the cost of the process is involved in the gasification steps. The costs depend on a supply of low-cost coal and are estimated in Tables 3-4, 3-6 and 3-8.

The process thermal efficiency is low and the slate of products produced includes many for which there is no large demand. The low-octane gasoline is of inferior quality and requires upgrading. Several direct liquefaction processes offer higher thermal efficiency, a product range that gives a better fit to the established market pattern and significantly lower costs. At least five years of additional development time will be required for expansion of the alternative technologies into commercial sized plants, while the F-T process is commercially available in large size plants today.

Table 3-9. Fischer-Tropsch Commercial Plant Mass Balance
(Stream-Day Basis for Comparison with Table 3-2)[a]

Coal Handling Facility Input (coal as mined)		27,792	tons
Process Input			
Moisture- and Ash-Free Coal (per Table 3-1)		18,593	tons
Gasifier Input Coal		15,264	tons
Boiler Input Coal		3,329	tons
Raw Water Input		39,840	tons
Raw Iron Ore (used as catalyst)		82.2	tons
Process Output			
Synthetic Pipeline Gas		173.3×10^6	scf
Gasoline		13,580	bbl
Diesel Fuel		2,307	bbl
Heavy Fuel Oil		622	bbl
Alcohols		1,825	bbl
Butane		146	bbl
Propane		1,107	bbl
Sulfur		61	tons
Anhydrous Ammonia		103	tons
Excess Electric Power		3.31	MW(e)
Thermal Conversion Factors			
Coal HHV	12,720 Btu/lb	7060 kg cal/kg	
Synthetic Gas	1003 Btu/scf	37.6 J/normal m^3	
Gasoline	5.0×10^6 Btu/bbl	5.28×10^6 kJ/bbl	
Heavy Hydrocarbons	5.5×10^6 Btu/bbl	5.80×10^6 kJ/bbl	
Butane and Propane	4.0×10^6 Btu/bbl	4.21×10^6 kJ/bbl	
Alcohols	3.8×10^6 Btu/bbl	4.0×10^6 kJ/bbl	
Thermal Efficiency			58%

[a]Source: Schreiner, 1978.

ZINC HALIDE HYDROCRACKING PROCESS

The zinc chloride catalyst process is designed to convert bituminous and subbituminous coal into distillates by catalytic hydrocracking in a bath of molten zinc chloride. The technical advantage of the process is that the catalyst converts a high percentage of the coal into gasoline and middle-range distillates. The process produces gasoline with a satisfactory octane rating and distillate oils that have lower sulfur and nitrogen content. The disadvantages are that the catalyst is a hot corrosive medium that is difficult to handle and requires regeneration. Development into an economical commercial process will require demonstration unit results that provide satisfactory solutions to current technical problems.

Process Description

Either coal or a solvent refined product may be utilized as a feedstock. The coal is dried and pulverized before mixing with a process derived recycle oil as illustrated in Figure 3-5. The slurry feed is fed into the hydrocracking reactor which consists of a bath of molten zinc chloride at a temperature of 752 to 842°F (400 to 450°C) and a pressure of 1500 to 3500 psig 105 to 246 kg/cm^2). The coal is cracked into gasoline and middle range distillates and is withdrawn into a receiver where the gas- and solid-free liquid products are separated by distillation. The spent catalyst contains the ash, nitrogen and sulfur compounds, and carbonaceous residue. It must be successfully regenerated if the process is to be economical. The catalyst is fed to a fluidized-bed combustor where the residual carbon is burned at 1700°F (912°C) to vaporize the catalyst, condense it and recycle it back to the reactor. The deficiency of recovered catalyst must be made up with expensive fresh zinc chloride.

Technical Evaluation

Project History and Status

The original work on the process was performed by Consolidation Coal Company between 1963 and 1968. After remaining inactive for several years the Energy Research and Development Administration (ERDA) revised the effort in 1975 to refurbish the bench scale units and resume testing. Following the demonstration of successful results

100 LIQUEFACTION

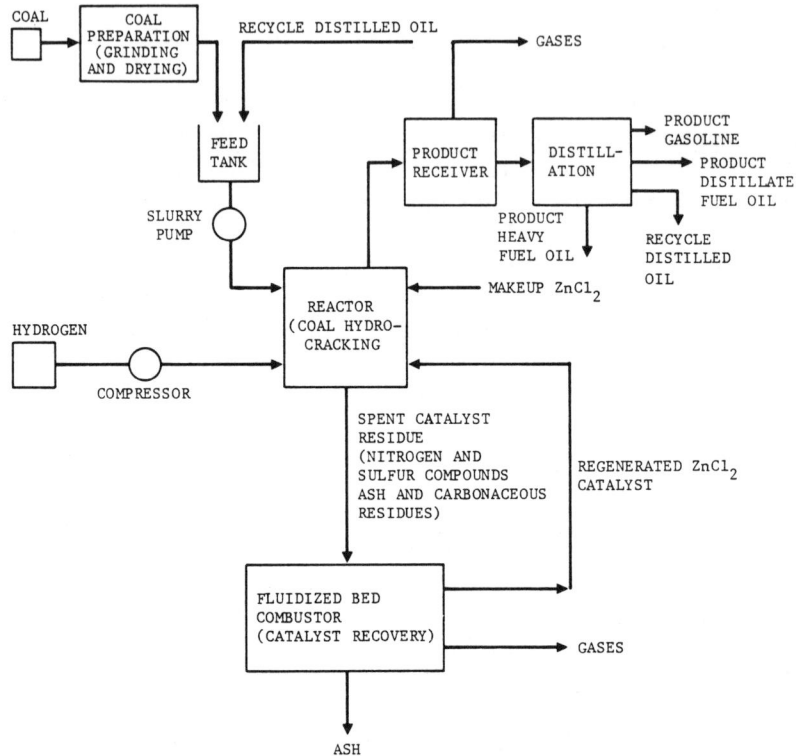

Figure 3-5. Zinc chloride catalyst process.

the design of a 1-ton/day development unit was begun in 1976. The unit was completed and began operation in 1978. It is scheduled to operate until the third quarter of 1980 on only one type of coal. The project results will then be assessed to determine if further development will occur.

The objective of the program is to develop a process producing clean liquid and gaseous fuels having high yield and distillate content. Of the product, 50% is supposed to be distillate, which is equivalent to producing 4 bbl/ton coal on a moisture- and ash-free basis. The work involves bench-scale studies and process development unit work to investigate the use of zinc halide as a Lewis acid catalyst for the hydrogenation and hydrocracking of subbituminous coals and of solvent refined coal extract. An economical regeneration process is to be developed for efficient recovery and recycle of zinc halide from the spent melt.

A continuous 2-lb/hr (0.908-kg/hr) bench-scale zinc halide hydrocracking unit and a 5-lb/hr capacity (2.27-kg/hr) continuous fluidized-bed combustion unit for regeneration of the spent zinc halide melt has been used to produce a low carbon spent melt. Solvent refined coal from Ft. Lewis, WA, was initially utilized to minimize operating problems. Process variables of temperature, pressure, space velocity and zinc chloride–to–coal ratio have been explored and used to develop a kinetic model. Distillates with boiling temperatures exceeding 392°F (200°C) and naphtha boiling at less than 392°F (200°C) were recycled through the reactor to measure stability and rates of conversion. Satisfactory operation with solvent refined coal and a bituminous coal was demonstrated.

The fluid-bed combustor was used to regenerate a low carbon spent melt. Zinc and halogen losses were evaluated and the recovery of zinc from the ash stream with the production of fuel gas was evaluated.

Following the bench-scale work, a 1-ton/day (0.9-metric-ton/day) hydrocracking and regeneration unit process development unit was constructed. Operations on the combined unit will provide information on the commercial potential for the process by studying the effect of equipment size on reaction rate, performance of mechanical components, suitability of construction materials and overall reliability. A 21-month operating period began in July 1978 when the reactor was completed. Problems were experienced with reliable operation of the hydrogen compressors and maintaining uniform temperature control in all parts of the reactor system. The reactor is operated at pressures up to 3500 psig (246 kg/cm^2) and temperatures up to 932°F (500°C). Under these conditions, zinc chloride becomes difficult to handle and causes corrosion in equipment and piping. Special corrosion-resistant alloys must be utilized for all metal components. Problems have been encountered in obtaining the predicted performance from pumps handling the molten zinc chloride.

The bench-scale tests have attempted to achieve operating conditions at which process conditions are optimized. Difficulties have been encountered in getting carbon content in the spent catalyst low enough for satisfactory regeneration of the spent melt. Regenerated catalyst has not been able to demonstrate the same reactivity as fresh catalyst in converting the higher boiling point oils to gasoline. The available evidence suggests that this is due to the buildup of chlorides (NaCl, KCl, CaCl$_2$, FeCl$_2$) in the spent melt resulting from reactions with the ash in the coal.

Process Development Unit Results

Representative results from the bench-scale tests illustrate the potential advantage of the zinc chloride process, a high conversion of coal

102 LIQUEFACTION

into gasoline and middle range distillate. Runs 46 and 49C were made with Kentucky No. 9 bituminous coal and its solvent-refined coal (SRC) extract at equal processing conditions. The percentage of carbon in the product as a percentage of the coal feed is a measure of the catalyst reactivity, since it is independent of the differing oxygen content of the feedstocks.

The data show that the SRC is substantially less reactive than its parent coal. Note that the conversion of the coal into the C_5-392°F (200°C) and 392-887°F (200-475°C) distillate range is almost 75%. Other runs with subbituminous coal and SRC have shown no reduction in reactivity when using the SRC.

Process Economics

An economic study has been conducted to estimate the process operations and costs of a 50,000-bbl/day gasoline plant using zinc chloride hydrocracking from Western coal and from solid SRC-I feedstock. Using processing parameters derived from the bench-scale work, material balances, flow sheets, equipment specifications and estimates of equipment costs will be developed. Using discounted cashflow as a parameter, the costs of production of gasoline from Colstrip, MT, coal, and from its SRC extract, will be estimated. The relative cost of gasoline from the two process variations will be more significant than the estimated absolute costs. This study was completed in 1978.

Assessment

Developmental support for the zinc halide process was terminated in 1979.

BIBLIOGRAPHY

Greene, C. R., et al. "Zinc Halide Hydrocracking Process for Distillate Fuels from Coal. Quarterly Technical Progress Report, August 1-October 31, 1978," Conoco Coal Development Company, Research Division, Library, PA (1978).

Lee, W., J. Maziuk, and W. K. Thiemann. "A New Process for Conversion of Coal to Gasoline," paper presented at the 26th DGMK Congress, Berlin, West Germany, October 1978.

Meisel, S. L. "Recent Advances in the Production of Fuels and Chemicals over Zeolite Catalysts," paper presented at the Leo Frend Symposium, American Chemical Society, Chicago, IL, August 1977.

Meisel, S. L., J. P. McCullough, C. J. Lechthaler and P. B. Weisz. "Gasoline from Methanol in One Step," *Chemtech* (February 1976).

Schreiner, M. "Research Guidance Studies to Assess Gasoline from Coal by Methanol-to-Gasoline and Sasol-Type Fischer-Tropsch Technologies, Final Report," FE-2447-13, Mobil Research and Development Corporation, Princeton, NJ (1978).

Struck, R. T., et al. "Zinc Halide Hydrocracking Process for Distillate Fuels from Coal, Quarterly Technical Progress Report, February 1-April 30, 1978," Conoco Coal Development Company, Research Division, Library, PA (1978).

Wise, J. J., and A. J. Silvestri. "Mobil Process Efficiently Converts Methanol to Gasoline," *Oil Gas J.* (November 1976).

CHAPTER 4
COMMON PROBLEMS IN LIQUEFACTION

ENVIRONMENTAL AND HEALTH PROBLEMS

An examination has been performed on the waste streams that are predicted to come from plants utilizing the major coal liquefaction technologies. Despite the technical differences between the processes, there is a strong similarity between the types of waste that will be produced and treatment technologies that will be required.

Established control technologies exist for most of the major waste components, such as sulfur dioxide, hydrogen sulfide, ammonia and phenols. However, they have not yet been proven in practice in the synthetic fuels industry and implemented on the large scale required to support a commercial plant operation. A major need exists for quantification of the discharges from existing pilot-plant operations so that the constituents and concentrations of the waste streams can be determined from actual measurements rather than concentrations based on mass balance calculations or inadequate measurements.

The limitations on the existing data that do exist from synthetic fuels plants have been summarized in a recent EPA report.

- The operation of pilot plants in the United States has been directed toward process development, with little emphasis on the environmental characterization of the process and waste streams. Processes have not been optimized to minimize environmental emissions or assess control technology needs.
- Only the dry-ash Lurgi and Sasol processes have been operated commercially abroad. These commercial plants do not incorporate many of the processes proposed for U.S. plants.
- Commercial plants operating in foreign countries often do not include pollution control equipment and utilize different types of coal than U.S. plants would consume.

- Although many unit operations for processing and pollution control have operated commercially in other industries, there are differences which may prove to be environmentally significant.
- Since some of the processes have been developed privately, the data are considered proprietary and are not publicly available.
- For many operations where discharge stream characterization data are available, the data are not complete. Not all streams have been characterized, and not all potential pollutants and toxic substances have been identified. Some data have not been obtained using the most modern analytical methods.
- Although the American development of coal conversion technologies is several decades old, interest in environmental assessment of the technologies is very recent.

The program that is recommended to overcome these deficiencies is based on a comprehensive set of multimedia measurements taken when integrated pilot plants begin operations. Comprehensive characterizations should be made on the toxicological and ecological characteristics of the waste streams. Detailed chemical analyses should be performed with emphasis on trace metals and organic chemicals. The performance of the control systems should be evaluated and related to engineering studies and the results of bench-scale testing. The results should be available in a timely fashion to (1) include environmental considerations in the selection of process equipment and waste management options for commercial plants, and (2) permit the drafting of new source performance standards (NSPS) for synthetic fuels facilities.

Because of the highly cyclic structure of coal, the products produced during liquefaction contain large quantities of polycyclic aromatic hydrocarbons (PAH). These substances have been recognized as the biologically active chemicals having carcinogenic, mutagenic and teratogenic effects. The effect of these substances on the health of the workers is a significant concern that is difficult to assess quantitatively before commercial scale plants are constructed. In addition to increased concentrations of aromatic and cyclic hydrocarbons in the liquid products, significant worker exposures may also occur due to leaks, fugitive emissions, process upsets, spills and maintenance activities. The worker exposure that will result from participation in these activities is unknown. Even where exposure levels are known, there frequently are inadequate health data to predict the effects. Existing knowledge is based on studying populations of coke oven workers and similar groups.

It must be stressed that no experience exists in the United States with coal liquefaction plants on a size that is considered of commercial scale, processing 25,000–50,000 ton/day of coal. All such plant designs are conceptual. Limited experience has been obtained with pilot plants that are a factor of 100 or more smaller in size. One may speak in principle of the processes and types of equipment that would be used to control environmental emissions and worker exposures. Relevant experience in other industries may be utilized where available. Until the first few plants of commercial size are designed, constructed, tested and operated for several years, no guarantee can be given that the proposed equipment and methodology will work as it is presently planned.

LIQUEFACTION PROCESS PROBLEMS

Commonly discussed problem areas in the four major liquefaction technologies include:

- mechanical equipment reliability,
- wastewater treatment,
- solid waste disposal,
- control of atmospheric emissions,
- processing of liquefaction bottoms,
- hydrogen production and recovery, and
- scaling upward in size of process steps and equipment.

Mechanical Equipment

Due to the abrasive nature of the coal-oil slurry that must be handled, pilot-plant operations have encountered significant problems in keeping mechanical equipment such as pumps, compressors and heat exchangers operating properly. Careful selection of equipment is required to withstand service in the high-temperature, high-pressure environment. Special selection of materials and hard surface coatings are often required. Pressure letdown valves have experienced high rates of wear, and filtering equipment often clogs or fails to operate when handling heavy, viscous oils. Scaling of pressure vessels and piping and the formation of carbonate deposits from calcium in the coal must be controlled adequately.

Wastewater Treatment

Numerous wastewater streams are generated from many steps in the processes of liquefaction. Representative contaminants to be treated include sour water containing oils, phenols, NH_3, H_2S, CO_2, SO_2, HCN,

108 LIQUEFACTION

NO_x and spent solvents. Blowdown will be generated from boilers and cooling towers. Sludges will result from raw water and wastewater treatment, and may also be produced by limestone flue gas scrubbers. The wastewater used to quench hot coal ash slag requires adequate treatment. Spent catalyst and contaminant absorbents must be disposed. Hydrocarbon vapors will leak, and spills of liquids will contaminate the water. Coal pile leachate runoff must be treated.

Solid Waste Disposal

A coal liquefaction facility will produce a large volume of solid wastes. Since high-ash coals are prime candidates for conversion, 15–20% of the raw coal may consist of inert material requiring environmentally acceptable disposal by landfilling. To the volume of ash material must be added the sludges produced by the various treatment processes. This may produce a mass of potentially hazardous material requiring suitable disposal that amounts to 25% of the coal that is processed.

The operation of a commercial sized synthetic liquid fuels plant produces large volumes of waste material that must be treated and properly disposed. A plant that will process 23,000 metric tons of coal per day will also consume 45,000 metric tons of raw water. For the H-coal process, the waste streams produced per day are estimated as:

- sour water: 5780 metric tons
- solids residue and char: 1480
- flue gases: 8190
- sludge: 1930
- acid gases: 10,130
- spent catalysts: 8
- particulates from coal preparation: 260

Although technologies exist for treatment of these wastes, the large volumes involved pose a problem of considerable magnitude. The area required for acceptable landfilling of toxic wastes so that they are isolated from contact with the biosphere due to leaching consumes several square kilometers over the 20- to 30-year lifetime of the plant. The total amount of land that will be affected by a commercial synthetic fuels plant, including the coal mine, waste disposal site, coal conversion plant, and water storage and treatment areas can total 200 mi^2 (520 km^2).

Atmospheric Emissions

Atmospheric emissions will contain the oxides of carbon, sulfur, nitrogen and particulates from many heaters that burn fossil fuel. This may be in the form of fuel gas produced in the process, liquid hydrocarbon fuel oils or raw coal burned to produce electric power and steam. Particulate emissions will be produced by coal handling and crushing operations, landfilling of solid wastes and construction activity. Adequate control measures will be required to limit hydrocarbon vapor emissions from product storage tanks. Significantly large transient emissions may be produced during startup, shutdown and process disturbances. Large-scale utilization of synthetic fuels derived from coal will increase atmospheric carbon dioxide concentrations which can lead to global climatic changes.

All of the processes employ commercial treatment processes, such as the Benfield, Purisol, Rectisol, Selexol or Sulfinol, to remove acid gases. The gases vented from the liquefaction product separation step and scrubbing liquors are treated to remove H_2S, COS, CS_2 and mercaptans. The sulfur is reclaimed in a sulfur recovery plant; unreacted hydrogen is recycled through the liquefaction reactor. The gases produced from gasification of the vacuum tower bottoms are also treated.

Liquefaction Bottoms

All of the coal liquefaction processes produce a heavy vacuum tower bottoms material from the product separation step. This material consists of unreacted coal, ash, heavy oils and other hydrocarbons. The optimum processing and disposition of this material has not yet been determined. Recycling through the reactor may increase the conversion efficiency of the process. Separation of the ash from the oil is a difficult processing step that has met with limited success, although many different technical approaches have been attempted. Subjecting the bottoms to pyrolysis drives off usable product oils and produces a char consisting of coke and ash. This char may be gasified to produce the hydrogen and fuel gas required by the process. If all of the organic material is not destroyed, disposal of the ash may be more critical because of the increased toxicity of the material. Optimum operation of a commercial plant would produce a material balance in which the process would be adjusted so that the bottoms produced would just suffice to meet plant requirements for hydrogen and fuel gas production. But process upsets,

110 LIQUEFACTION

unexpected coal properties or variations in operating parameters could produce more bottoms material requiring disposal than can be utilized in the process. A potential utilization for the bottoms is as a high-ash fuel for a plant boiler. Several types of liquefaction residues have successfully been gasified in an entrained bed gasifier pilot plant.

A review of the liquefaction processes has shown a number of common processing steps that are encountered in most of the technologies. Improved solutions to any of the problems listed in Table 4-1 will result in improvements for all of the processes.

Table 4-1. Common Liquefaction Process Steps

Coal Handling and Treatment
    Crushing and grinding
    Sizing
    Drying
Pyrolysis of Heavy Oils
Catalytic Synthesis of Liquid Hydrocarbons
Separation of Solids, Liquids and Gases
    Flashing and condensation
    Filtering
    Centrifuging
    Solvent de-ashing
    Vacuum distillation
    Coking
    Fractionating
    Hydroclones
    Cyclone
Hydrotreating to Remove Sulfur, Oxygen and Nitrogen
Hydrogen and Synthesis Gas Generation
Oxygen Generation
Acid Gas Removal
    CO_2
    H_2S
    SO_2
    Organic sulfur
Sulfur Recovery
Product Storage and Transportation
    Sulfur
    Anhydrous ammonia
    Organic chemicals
    Gasoline
    Fuel Oil
    Distillates
    Liquefied petroleum gas
Raw Water Treatment

Table 4-1, continued

Wastewater Treatment
   NH_3
   H_2S
   Tars
   Phenols
   Liquid hydrocarbons
   Organic sulfur compounds
Gaseous Waste Treatment
   NH_3
   H_2S
   SO_2
   NO_x
   Particulates
   Polycyclic organic material
Solid Waste Disposal
   Coal ash
   Spent catalyst
   Scrubber sludges
   Spent plant chemicals
   Water treatment sludges
Particulate Material Collection
Transient Spill Control
Miscellaneous By-product Recovery
   Sulfur
   Ammonia
   Phenol
   Tar and oils
   Light liquid hydrocarbons
   Char
   Fuel gases
Steam and Power Generation
   High- and low-pressure steam
   Electric power
   Waste heat recovery
   Cooling water

Cost of Hydrotreating Coal Liquefaction Products

The cost of hydrotreatment of coal liquids to produce a fuel suitable for use in gas turbines was estimated by Mobil research and development (R&D). The study was performed for the Electric Power Research Institute (EPRI) and was intended for utilization in an electric utility environment. Five of the liquids produced by American liquefication processes were considered as feedstock. The liquids differ greatly in their initial hydrogen composition and ash content. The processing costs varied with the severity of hydroprocessing required, which was a function of feedstock quality and product quality.

The study assumes that the hydrotreatment plant would be constructed as part of a petroleum refinery. Hydrogen would be available from the refinery at \$2.00/1000 scf (\$2.00/2832 cm^3), as well as other utility services such as electric power and steam. The facility size would be 20,000 bbl/day of feed material, with other components sized to meet the requirements of each processing option.

The major processing sections of the plant are the hydrotreater, sulfur recovery plant and wastewater treatment to recovery ammonia. For distillate stocks, a fixed-bed reactor design was chosen, using multistage reactors with multiple quench zones. The estimates of catalyst performance, product yields and product properties were based on experimental data employing a commercial CoMo on alumina catalyst. For residual stocks, ebullated-bed reactors were capable of continuous replacement of the large-pore NiMo on alumina catalyst. The processes are based on Mobil experimental, bench-scale data on coal liquid upgrading. The sulfur recovery and water treatment plants use commercial technology.

The five feedstocks are:

1. solvent-refined coal (SRC) light organic liquid,
2. H-coal distillate,
3. SRC recycle solvent,
4. H-coal 400°F+ (204°C+) fuel oil, and
5. solvent refined coal.

The SRC light organic liquid boils in the 150–425°F (55–218°C) range and has a high heteroatom content, but no ash. H-Coal distillate boils in the 325–700°F (163–371°C) range and is made in the fuel oil mode, with the lowest heteroatom content and no ash. SRC recycle solvent is in the same 325–700°F (163–371°C) boiling range, but represents a poorer feedstock in terms of hydrogen and heteroatom content. H-Coal 400°F+ (204°C+) fuel oil is a heavy residual product from which the solids and unreacted coal have been removed by solvent precipitation. It is similar to SRC recycle solvent, but has high-molecular-weight asphaltic material and a different heteroatom content. It contains 200 ppm of ash, which will rapidly deactivate the catalyst. Solvent-refined coal requires the most severe processing conditions due to the low hydrogen content and high ash content (1000 ppm).

COMMON PROBLEMS IN LIQUEFACTION

A total of 15 processing configurations were evaluated, representing a range of processing severity, producing turbine fuel containing from 8 to 13% hydrogen by weight. The hydrotreater, sulfur plant and wastewater treatment facilities were optimized for each processing configuration. Catalyst consumption, hydrogen and other chemicals, steam, electric power, water, fuel, and labor costs were determined. The processing conditions were varied, and yields of hydrogen, ammonia, hydrogen sulfide, water, and light and heavy hydrocarbons were determined. The gravity, heating value, elemental and trace element composition, and ash content of the product liquid were determined. By-product credit was given for ammonia, but not for sulfur. Hydrogen and hydrocarbon offgases were credited for use as a fuel gas.

The economic analysis calculated investment costs for the facility, utilities, general support facilities and chemical costs. Project contingency costs were included, as were taxes, preproduction costs, construction loan interest, working capital, utilities, labor, management, maintenance, properties and insurance. Annual capital charges were determined by both utility and equity financing using a discounted cashflow rate of return on equity.

Table 4-2 gives the resulting costs for the middle hydrogen content product for each of the five feedstocks. About 50% of the total cost is for hydrogen input; the other 50% breaks down about equally into

Table 4-2. Cost of Upgrading Coal Liquids to Turbine Fuels
(Equity Financing, 5%/year Inflation)[a]

	SRC Light Organic Liquid	H-Coal Distillate	SRC Recycle Solvent	H-Coal 400°F+ Fuel Oil	Solvent-Refined Coal
Hydrogen Content (wt %)					
Feedstock	11.3	10.1	7.7	7.4	5.7
C_5+ Product	12.0	11.4	10.0	10.0	9.0
Sulfur Content (wt %)					
Feedstock	0.06	0.11	0.02	0.48	0.6
C_5+ Product	0.08	<0.01	0.01	0.02	0.15
Ash Content (ppm)					
Feedstock				200	1000
C_5+ Product				20	120
Capital Investment ($ Million, 1980)	51	94	108	136	182
Net Cost, 1980 ($/$10^6$ Btu)	0.52	0.93	1.36	1.53	2.02

[a] Source: EPRI AF-710.

25% for operating cost and 25% for capital recovery. The size of the hydrotreatment facility increases greatly as the required processing becomes more severe. Total costs vary from $0.52 to $2.02 per 10^6 Btu (251,980 kg-cal) for 1980 equity financing, with required capital investments of $51 million to $182 million.

ECONOMIC ANALYSIS OF COAL LIQUEFACTION PROCESSES

A technical and economic review of the principal coal liquefaction processes was undertaken in 1978 by an organization independent of the developers and promoters of the technology. The authors of the Engineering Societies Committee on Energy, Inc. prepared a report for the U.S. Department of Energy (DOE) in which conceptual commercial-scale plants were evaluated by means of an objective cost index. This cost index reflected plant investment, coal cost, labor and utility costs, and selling prices for each of the products produced. The cost of a standard-size plant was estimated for all processes on a consistent basis. The resulting cost index has comparative value only and is not intended as a prediction of prices.

The cost index is an incomplete hypothetical average product cost. Many real costs are not included in the computation of the cost index. Taxes, actual interest rates, future inflation, working capital depreciation, price regulations and government-provided incentives will make significant contributions to the real prices of synthetic liquid fuels, but are not included in the cost index. The cost index does consider the equipment demands of the process and the desirability in the marketplace of the fuel that it produces. The index does include labor cost, coal, 12% interest, a plant lifetime of 20 years and utility requirements.

The quantities of different output products produced by the seven processes and operating modes are shown in Table 4-3. To compare the outputs two relative value factors were adopted. The 1978 value factor represents the price distribution when all forms of petroleum products were commanding high prices. The 1970 value factor represents prices when residual fuels were available in sufficient quantities to depress prices relative to the fuel in most demand, automotive gasoline. The product value factors used in the analysis are given in Table 4-4. A considerable increase in fuel oil value relative to gasoline from 1970 to 1978 is noteworthy.

The plant capital investment is an important contributor to calculating the cost index. Because the published cost estimates for each process came from different sources and were done in different time periods, adjustments were made to place the estimates on the same consistent basis. All of the processes utilize commercial technology on a scale signifi-

Table 4-3. Products and Plant Capital Cost[a]

Process Operating Mode	Products (bbl/stream-day)	Plant Capital Cost ($ million, 1978)
Fischer-Tropsch	19,600 gasoline	688
	20,300 LPG	
	1,300 No. 2 fuel oil	
	2,100 No. 6 fuel oil	
Mobil-Gasoline from Methanol	47,800 premium gasoline	794
	5,700 LPG	
H-Coal (Syncrude Mode)	24,700 naphtha	696
	36,400 syncrude	
H-Coal (Fuel Oil Mode)	15,500 naphtha	586
	51,300 No. 6 fuel oil	
Exxon Donor Solvent	27,500 naphtha	779
	10,700 LPG	
	37,200 No. 6 fuel oil	
Solvent Refined Coal-I	13,000 naphtha	740
	64,400 solid (equivalent)[b]	
Solvent Refined Coal-II	13,000 naphtha	774
	6,400 No. 2 fuel oil	
	52,900 No. 6 fuel oil	

[a]Source: Rogers et al., 1978.
[b]SRC-I equivalent based on 6.2 million Btu/BBL.

Table 4-4. Product Value Factors[a,b]

Product	1978 $/bbl	Value Factor 1978	Value Factor 1970
Premium Gasoline	17.50 = A	1.0	1.0
No. 6 Fuel Oil	12.30 = B	0.70[c]	0.44
SRC-I Solid		0.63	0.40
No. 2 Fuel Oil	14.90 = C	0.85[d]	0.71
Naphtha			
H-Coal		0.89	0.79
Exxon Donor Solvent		0.88	0.78
Solvent-Refined Coal		0.85	0.71
LPG	12.12	0.69	0.43
Regular Gasoline	16.30	0.93	0.87
Fischer-Tropsch Gasoline		0.90	0.84

[a]H-Syncrude is equivalent to No. 3 fuel-oil; SRC and EDS fuel oils are No. 6 fuel.
[b]Source: EPA-60017-78-168a.
[c]B/A.
[d]C/A.

116 LIQUEFACTION

cantly larger than that which has been constructed at the present time. Consequently, the forecasting of capital costs is difficult to do accurately, but since the processes utilize many common elements, the forecast errors should affect each process equally.

Estimates were made for the costs of comparable plants consuming 25,000 tons (22,695 metric tons) per stream-day* of dry Illinois #6 coal at a Gulf Coast location. The plant capital costs are shown in Table 4-3 as well as the quantities of output product produced from each process. The same labor cost was used for each plant. Delivered coal was priced at $15/dry ton. The cost of utilities and credit for surplus electric power produced was included. Hydrotreatment costs for upgrading naphtha to gasoline are included.

Table 4-5. Process Comparison[a]

Process	Liquid Products Thermal Efficiency[b]	Cost Index[c] 1978	1970
Fischer-Tropsch	32%	20.5	16.0
Mobil Gasoline	44	14.0	14.5
H-Coal Syncrude Mode	56	12.6	14.7
H-Coal Fuel Oil Mode	66	12.9	18.4
Exxon Donor Solvent	65	13.6	18.5
Solvent Refined Coal-I[d]	70	13.6	20.1
Solvent Refined Coal-II	77	13.7	20.4

[a]Source: Rogers et al., 1978.
[b]Only liquid products are counted in thermal efficiency calculation. Solid and gaseous products are not counted.
[c]Refer to cost index definition in text.
[d]SRC-I solid treated as a liquid for this purpose.

The process costs were used to calculate the cost index for each case using the relative product value factors. The calculated cost index for each process is given in Table 4-5. It must be emphasized that these numbers are derived only to compare processes on a consistent relative basis and are not valid price predictions. All process heat is not accounted for and the by-product values of sulfur and ammonia are not included.

Examination of the results shows that the Fischer-Tropsch process has a significantly higher cost than the others. The other six processes can be compared with each other. When the analysis is performed for the 1978 product values, the cost index difference for the six processes

---

*A year contains 360 idealized stream-days. Total yearly production at 90% capacity factor is 0.9 × 360 day/idealized year = 324 stream-days.

is less than 10%. Since the size of the conceptual commercial plants is at least a factor of 100 greater than any pilot plant now operating, the uncertainties in the method of estimate exceed any prediction of the relative cost of product.

The study concludes that any of the processes can be made to operate at a commercial scale. On the basis of existing development work it may be anticipated that problem areas will arise that will alter the economics on a commercial scale. But these cannot be accurately forecast at present, even though some contingency funds have been included for processes where additional development is needed. The processes produce a wide range of products from a solid substitute for coal through the fuel oil and distillate range to gasoline and liquefied petroleum gas. Given the uncertainties in predicting plant costs, process development uncertainties, relative product values and the variations caused by geographical location, price of coal, environmental control costs and value of excess fuel gas and electric power, there is no reason to choose between any of the six technologies based on our ability to predict a cost advantage. Only the Fischer-Tropsch technology has a definite cost disadvantage, as well as a low thermal efficiency.

BIBLIOGRAPHY

Mobil Research and Development Corp. "Upgrading of Coal Liquids for Use as Power Generation Fuels," report to the Electric Power Research Institute, EPRI 361-1 (1976).

Rogers, K. A., A. S. Wilk, B. C. McBeath and R. F. Hill. "Comparison of Coal Liquefaction Processes," The Engineering Societies Commission on Energy, Inc., Washington, DC (1978).

TRW Environmental Engineering Division. "Environmental Assessment Data Base for High Btu Gasification Technology," EPA-60017-78-168a (1978).

SECTION 2

GASIFICATION

CHAPTER 5
INTRODUCTION

GASIFICATION PROCESSES

Although coal gasification has been practiced for more than a century, the engineering design of the processes used is mostly empirical. The most commonly used gasifiers are classified as fixed-bed, entrained-bed and fluid-bed by process type. A more theoretical classification considers the range of temperatures under which the gas is formed and the quantity of steam used per unit weight of coal. A gasification process must satisfy chemical constraints based on the stoichiometry of the coal gasification reactions and the energy requirements to produce those reactions. Low-temperature processes are thermally efficient and maximize methane formation, but have low throughput and produce tars. Higher-temperature processes avoid these problems, but must recover more of the coal energy in a heat recovery boiler. Process efficiency is improved by using smaller amounts of steam, but these processes have proven more difficult to operate. High steam usage produces a high ratio of hydrogen to carbon monoxide in the output gas. The process selected must be matched to its intended usage.

Calculations performed in modeling reactions used to produce low-Btu gas have increased our understanding of the processes involved. The basic understanding is at an elementary level because the chemical reactions are not well understood, and the rate constants are unknown and vary with the rank of coal, the catalytic activity of the mineral matter and pretreatment methods used. In a fixed-bed gasifier the feed coal is slowly warmed by the product gas stream and undergoes drying and devolatilization. Significant methane production occurs before the char is oxidized with air and reacted with steam. The heating value of the gas that is produced will increase to 150 Btu/scf (and slightly higher) as the gasification temperature is increased to about 1700°F (920°K).

The conversion of carbon to CO and production of H_2 increases as gasification temperature rises, while the production of CH_4, H_2O and CO_2 decreases. The production of trace elements and minor chemical species varies with temperature in a complete fashion.

SUMMARY OF GASIFICATION PROCESSES

The five gasification processes described in this book differ in mode of operation and characteristics of the products produced. Although the process developers have suggested numerous applications for the units, a careful study is necessary to select the optimum gasifier for a specific application and type of coal available as feedstock. Each gasifier has been developed from a specific historical context; any attempt to utilize the process for a different application should be analyzed thoroughly.

The slagging Lurgi process is primarily a methane and fuel gas producer, with the higher-temperature operation given throughput and thermal efficiency advantages over the established dry-ash Lurgi process. By reducing the steam injection requirements, the capital investment in equipment is reduced, and the process thermal efficiency is increased. The pressurized gasifier, operating at a higher temperature, produces a larger quantity of gas containing a higher concentration of methane.

The Texaco and Shell-Koppers processes were developed from petroleum refinery experience to produce a chemical synthesis gas. The unique Texaco water slurry injection system produces a high hydrogen–to–carbon monoxide ratio in the product gas. The Shell-Koppers process, a pressurized version of the Koppers-Totzek process, uses the minimum amount of steam for high thermal efficiency and produces a low hydrogen–to–carbon monoxide ratio. The intended application of the gas can determine the preferred gas composition. Both the Texaco and Shell-Koppers units require a high-temperature heat exchanger that represents a difficult technical problem, and has resulted in maintenance problems. Both processes give about half of the available energy in the form of steam from the waste heat boiler, which is easily utilized in a refinery or chemical plant, but may not have a use in other applications.

The Combustion Engineering (CE) two-stage gasifier produces at a lower temperature and utilizes a heat exchanger design like a utility boiler. Because of the atmospheric pressure operation and a design that gives quick access for maintenance of the components, unit availability should be high. The CE gasifier is designed as a high-volume producer of low-Btu gas to be fed directly into a utility boiler.

INTRODUCTION

The COGAS process was designed to utilize the char produced from the COED pyrolysis process for producing liquids. The low-temperature process gives high thermal efficiency and is the only one utilizing the fluid-bed principle. The gas is reformed into synthetic pipeline gas.

Table 5-1 summarizes some of the significant characteristics of the gasification processes. The selection of the optimum process for a given application requires a careful analysis in which the characteristics of the gasifiers are often designed to fit the application and type of coal used. Many potential applications prefer to avoid tar and oil production because of the difficulty of removing this material from the gas and finding a suitable usage. Steam produced from high-temperature raw gas may have a use, depending on the application. The chemical composition of the gas produced varies with the types of coal, gasifier conditions and design of the cleanup system. The gasifier pressure should be chosen to match the application.

Process data are summarized in Table 5-2. The construction and operating economics have been summarized in Table 5-3 with an attempt made to utilize a consistent method of reporting. Because of differences in the developers' methods of computation, comparisons between processes may not always be valid.

GAS CLEANUP SYSTEMS

The raw product gas that is produced is not suited for direct combustion under modern environmental standards. The hot gas contains coal dust, ash, char, HCl, NH_3, H_2S and other sulfur compounds, the vapors of alkali and other metals, and heavy polycyclic and heterocyclic aromatic oils known as tar.

The control technology that is employed must reduce the concentrations of all of the above contaminants to specifications that may become more stringent in future years. Although several attempts have been made to develop hot gas cleanup and desulfurization systems, they are not considered practical. The present sequence of treatments is the accepted practice in the United States:

1. recover usable heat in a heat exchanger;
2. utilize a water spray to remove NH_3, particulates and metal vapors; and
3. remove H_2S and other sulfur compounds with commercial desulfurization processes developed in the petroleum and chemical industries.

124 GASIFICATION

Table 5-1. Characteristics of Gasification Processes

Process	Gasifier Type	Product Gas Temperature	Tar and Oil Production	H2/CO Ratio	Application
Slagging Lurgi	Moving bed	Low	Yes	Medium	Methane and fuel gas
Texaco	Entrained bed	High	No	High	Chemical synthesis gas
Shell-Koppers	Entrained bed	Highest because steam injection is limited	No	Lowest	Chemical synthesis gas
Combustion Engineering	2-stage entrained bed	Medium	No	Medium	Low BTU atmospheric pressure fuel gas for utility boiler
COGAS	Multiple fluidized beds with char recycle	Low	Liquid fuel production is an objective	High	Coproduction of liquid fuels and synthetic pipeline gas

INTRODUCTION 125

Table 5-2. Gasification Process Data Summary

Process	Slagging Lurgi	Texaco	Combustion Engineering	Shell-Koppers	COGAS
Name of Coal	Illinois #6	NA	Pittsburgh #8	Good Quality	Illinois #6
Type of Coal	Bituminous	NA	Bituminous	Bituminous	Bituminous
Coal Composition	Table 2-4		Table 6-48	NA	Table 3-5
Material Balance (stream-day basis)					
Coal in	27,301 tons		4,800 tons	NA	25,935 tons
Electric Power in			NA		4.4 MW(e)
Raw Water in	73,234 tons				
Synthetic Gas	242 × 10⁶ scfd				16.5 × 10⁶ gal
Propane	—				265 × 10⁶ scf
Butane	—				—
Naphtha	180 tons				—
Fuel Oil	262 tons				3,915 bbl
Sulfur	918 tons				16,823 bbl
Ammonia	48 tons				682 tons
Phenol	68 tons				48 tons
Electric Power Out			574.6 MWe	NA	20.6 MW(e)
Energy Balance	Table 6-4		Table 6-47	Table 7-4	NA
Thermal Efficiency	59%		81%	77–80%	68%

126 GASIFICATION

Table 5-3. Gasification Process Economic Analysis

	Slagging Lurgi	Texaco	Combustion Engineering	Shell-Koppers	COGAS
Capital Cost ($ million)					
Construction	975	NA	389,580	NA	1,276
Catalysts and Chemicals	14				13
Startup Costs	57				39
Capacity Factor	90%		80%	NA	NA
Operating Life	20 yrs		NA	NA	NA
Construction Time	4 yrs		NA	NA	NA
Annual Operating Cost ($ million)					
Coal	$2.47 per 10^6 Btu	NA	77,752	NA	210
Other Raw Materials	0.08				2.5
Catalysts and Chemicals	0.20				23
Labor and Benefits	0.29		3,691		25
Maintenance Supplies	0.14		2,149		33
Maintenance Labor					
Other	0.58		778		10
By-product Sales	(0.93)	NA	640	NA	(41)
Required Personnel			195		NA
Operation					
Maintenance					
Administrative					

The purification of coal produced gas requires a control technology that must reduce the concentrations of particulates in the form of char and ash, corrosive compounds of chlorine, alkali metals, sulfates, other sulfur and nitrogen compounds, and organic polycyclic and heterocyclic aromatic hydrocarbons (tars) that can amount to 6% of the carbon in the coal. The specifications to be met by the clean gas vary with the application for which the gas is used. Chemical processing requires the highest level of cleanliness to avoid catalyst poisoning. Fuel gas specifications are often determined by environmental regulations.

Because there is no available commercial process for cleaning the gas at the exit temperature from the gasifier, U.S. practice requires cooling the gas to 50–129°C before further treatment. In many applications, the heat can be recovered for process application. In return for avoiding the technical problems of hot gas cleanup, a penalty is extracted in thermal efficiency amounting to 3% for some electricity generating cycles, to 8% for fuel gas applications. However, the simplicity and effectiveness of a water quench in a venturi or contact tower for removing ash and char particles, tars and oils, HCN, NH_3 HCl and chloride salts, and alkali metals is unequaled.

Sulfur is then removed using commercial processes developed in the petroleum and chemical industries. Adsorption onto solids in the form of molecular sieves, activated carbon, and iron and zinc oxides is limited to input concentrations of H_2S less than 1%, but can reduce concentrations in the product gas to less than 1 ppm. Adsorption onto solids can be used as a secondary treatment following one of the physical or chemical adsorption processes described below.

The Stretford process is widely employed in Europe for coke oven, refinery, synthesis and natural gases, and hydrocarbon liquids. Following absorption in an alkaline solution, an oxidation-reduction reaction is used to produce elemental sulfur. Process economics are best when capacity is less than 10 tons of sulfur per day, and when the H_2S concentration is small and the pressure is low.

Alkanolamines have been used in treating natural gas and refinery vapor streams for more than 40 years. Several of the amines can be used in a reversible adsorption-desorption chemical reaction with reasonable energy requirements. Some solutions irreversibly react with COS and CS_2; others will remove both CO_2 and H_2S. Sulfinol is a mixture of the physical solvent sulfolene and diisopropanol amine. The system works best at low partial pressures of H_2S and CO_2, and can produce a low residual concentration. Continual solution reclaiming is required because of side reactions with impurities such as carboxylic acids and HCN.

High partial pressures of H_2S and CO_2 in the feed gas favor physical absorption in solvents such as Rectisol which is a refrigerated methanol solution used for obtaining low H_2S concentrations. Fluor solvent, Purisol, Selexol and Estosolvan are four commercial processes. Selexol, for example, uses the dimethyl ether of polyethylene glycol and is favored at high pressures and low temperature. The solvent is stable, has low losses, is noncorrosive, and the carryover is not harmful to gas turbines. There are no reactions with COS, CS_2, NH_3 or HCN which are passed on to the Claus plant. The pickup of CO_2 is low, while removing H_2S down to 1 ppm.

The Benfield system, using hot K_2CO_3 to remove CO_2, H_2S, COS and CS_2, is economical and well suited to low- and medium-Btu fuel gas requirements. The process becomes uneconomical beyond 90–95% sulfur removal, which fits well with environmental requirements.

A Claus plant is the usual means of converting the adsorbed gas into elemental sulfur. Environmental regulations are met by treating the tail gas from the Claus plant by the Beavon, SCOT or IPS (Institute Français du Petrole) processes.

MATHEMATICAL MODELING OF GASIFIER PERFORMANCE

Stillman (1979) described a computerized simulation of an adiabatic steady-state plug-flow moving bed coal gasifier, serving the dual purpose of describing the theoretical simulation of a gasifier, and the physical processes and chemical reactions that occurred. The moving bed gasifier consists of a vertical cylindrical reaction vessel in which the coal is fed into the top and flows slowly downward while the gas stream flows countercurrent upward. Air or oxygen and steam are fed into the bottom, while ash or slag is removed. The gasified coal, entrained dust and contaminants exit the reactor at the top. Since the downward motion of the solids is very slow, another name often used is the fixed-bed reactor.

The model considers 17 solid stream components, 10 gas stream components and 17 chemical reactions. It has been applied to experimental results from a Wyoming subbituminous coal for which kinetic and thermodynamic coefficients were determined.

As the coal proceeds through the reaction vessel, the following physical and chemical events occur.

Drying

As the moist coal increases in temperature, water is the first constituent to evolve.

$$\text{moist coal} \rightarrow \text{dry coal} + H_2O$$

Devolatilization

As the temperature of the dry coal increases pyrolysis occurs.

$$\text{dry coal} \rightarrow \text{char} + \text{volatiles}$$

The 10 gases included in the model are H_2O, H_2, N_2, O_2, CO_2, CO, CH_4, H_2S, NH_3, and tar; the 17 components of the downward moving solids stream are C, S, ash, slag, clinker, H_2O, H_2, N_2, O_2; with the following components also considered as volatile solids: H_2O, H_2, CO_2, CO, CH_4, H_2S, NH_3 and tar.

Gasification

The gasification is represented by the following five chemical reactions.

$$\text{char} + H_2O \rightarrow CO + H_2$$
$$\text{char} + CO_2 \rightarrow 2CO$$
$$\text{char} + 2H_2 \rightarrow CH_4$$
$$CO + H_2O \rightarrow CO_2 + H_2$$
$$CO + 3H_2 \rightarrow CH_4 + H_2O$$

Combustion

The thermal energy which drives the reaction is provided by combustion of the char at the bottom of the reactor vessel producing ash which can be melted into liquid slag, or resolidified into clinker.

$$\text{char} + O_2 \rightarrow \text{ash} + CO + CO_2$$
$$\text{ash} \rightarrow \text{slag}$$
$$\text{slag} \rightarrow \text{clinker}$$

Kinetic Equations

Only a small part of the information describing the kinetic equations used in Stillman's model can be noted here. Devolatilization reaction rates were determined from pyrolysis experiments performed at a slow rate of temperature increase characteristic of the moving bed gasifier. The gasification model includes a factor for the change in pore surface area during the reaction and a change in the diffusional resistance within the solid particle as reaction temperature increases. The reactions between char and H_2O, CO_2 and H_2 are described by irreversible heterogenous gas-solid reactions. Changes in the effective surface area ratio as the char is consumed are included. The water gas shift reaction and methanation reactions have been adjusted to a thermodynamically consistent basis with the mineral content of the ash acting as a catalyst. The kinetic model of combustion utilizes a shrinking core model with the coefficients selected from data on the Wyoming subbituminous coal that was used in the experimental gasification trials. Rate constants were calculated from a first-order equation describing an irreversible gas-solid reaction considering the solid particle diameter, film mass transfer coefficient and diffusion of oxygen through the ash layer formed as the particle is consumed. Ash melting and clinker formation were modeled from simple heat transfer and temperature relationships.

Steady-State Differential Equations

The mathematical model for the reactor is obtained by setting up differential equations for each chemical constituent from the continuity equations for mass and energy. The concentrations of the constituents vary only along the vertical axis of the gasifier; no radical variation is considered and the reactor is assumed to be operating adiabatically. A total of 17 equations were set up for the mass balance of the solids streams, 10 equations for the gas streams and 27 additional equations for the steady-state energy balance of the solids and gases. Two additional equations were used to define a linear pressure drop across the length of the reactor and to define the density of the solids. Boundary conditions were established at the top and bottom of the reactor where the temperature, pressure and concentrations of the solids and gases in the feed and product material were set to fixed values.

The steady state differential equations form a 29th-order system that must be iterated to reach a solution matching the boundary conditions at

INTRODUCTION 131

the top and bottom of the reactors. The solution is begun by guessing at the composition of the gas leaving the reactor when feeding in the solid constituents known to be entering the reactor. The 29 differential equations are integrated from the top to the bottom of the reactor using a variable step Runge-Kutta-Fehlberg method. If the calculated gases entering the reactor do not match the known values, new exit gas constituents are assumed and the calculation is repeated. The procedure is implemented with a computer program that has been used to model a Lurgi reactor (dry ash), a slagging Lurgi reactor (molten slag), and an intermediate case (ash and clinker).

Comparison with Test Data

Roland seam subbituminous western coal has been used in the slagging Lurgi reactor at Westfield, Scotland. A comparison of the model results with test data has shown good agreement in the calculated temperature profile along the length of the reactor, but a slight shift in the location of the maximum temperature. An increase in the methanation rate reaction by a factor of ten was necessary to match the data. The model was then exercised to determine the effect of varying the amount of moisture in the feed coal. The residence time of the coal in the reactor increased because of the additional drying time required and the throughput decreased. The temperature profile of the reactor and the position of the four zones was changed. The effect of altering the temperature of the inlet gas and steam was computed to determine the change in the composition of the outlet gas. Finally the effects of processing the same Roland seam coal with the three different types of reactors was computed (Table 5-4).

As the reactor conditions progress from dry ash to slagging, the H_2 and CO_2 content of the exit gases is reduced, CO and CH_4 content increases. The exit gas temperature for the slagging reactors is lower because of the reduced gas flow, but the throughput is higher because of

Table 5-4. Effects of Processing Roland Seam Coal in Three Different Reactors

Reactor Type	Dry Coal/Oxygen (wt/wt)	Steam/Oxygen (mol/mol)
Dry Ash Lurgi	2.80	8.20
Clinker Forming	2.64	4.15
Slagging Lurgi	2.50	1.10

the more rapid rates at which the coal devolatilization, gasification and char combustion occur at the higher operating temperature.

BIBLIOGRAPHY

"Assessment of Low and Intermediate Btu Gasification of Coal," FE/1216-4, National Academy of Science, Washington, DC (1977).

Stillman, R. "Simulation of a Moving Bed Gasifier for a Western Coal," *IBM J. Res. Devel.* 23:240 (1979).

CHAPTER 6

MAJOR GASIFICATION PROCESSES

SLAGGING LURGI

Process and Manufacture

Process Description

The slagging Lurgi* gasification process manufactures substitute natural gas (SNG) from caking and noncaking coals, with high efficiency and a minimum of steam consumption. The technical feasibility of the process has been demonstrated in a series of runs conducted in the British Gas Corporation (BGC) 300-ton/day (270-metric ton/day) pilot plant located at Westfield, Scotland. In addition to the gasifier, a shift converter, gas cooler, purification methanation, drying and compressing units must be included in a well integrated system (Figure 6-1) to produce the output product. The gasifier used in this process is a pressurized, oxygen-blown, fixed-bed Lurgi dry-ash gasifier which has been modified to operate at temperatures above the clinkering point of most coal ashes (1500°C). At this temperature the influence of the coal's reactivity is negligible and it is possible to operate with steam-to-oxygen ratios ranging from 1.1 to 1.5 mol/mol.

The slagging Lurgi differs from the conventional Lurgi by allowing the ash to melt before it is collected at the bottom of the unit. The dry-ash Lurgi works with steam/oxygen ratios of 8 to 10 mol/mol. The excess steam is needed to cool the ash down well below its melting point to avoid clinkering effects and thus clogging of the slag retrieval mechanism.

---

*Proprietary technology held by Lurgi Kohle und Mineraloeltechnik and British Gas Corporation.

134 GASIFICATION

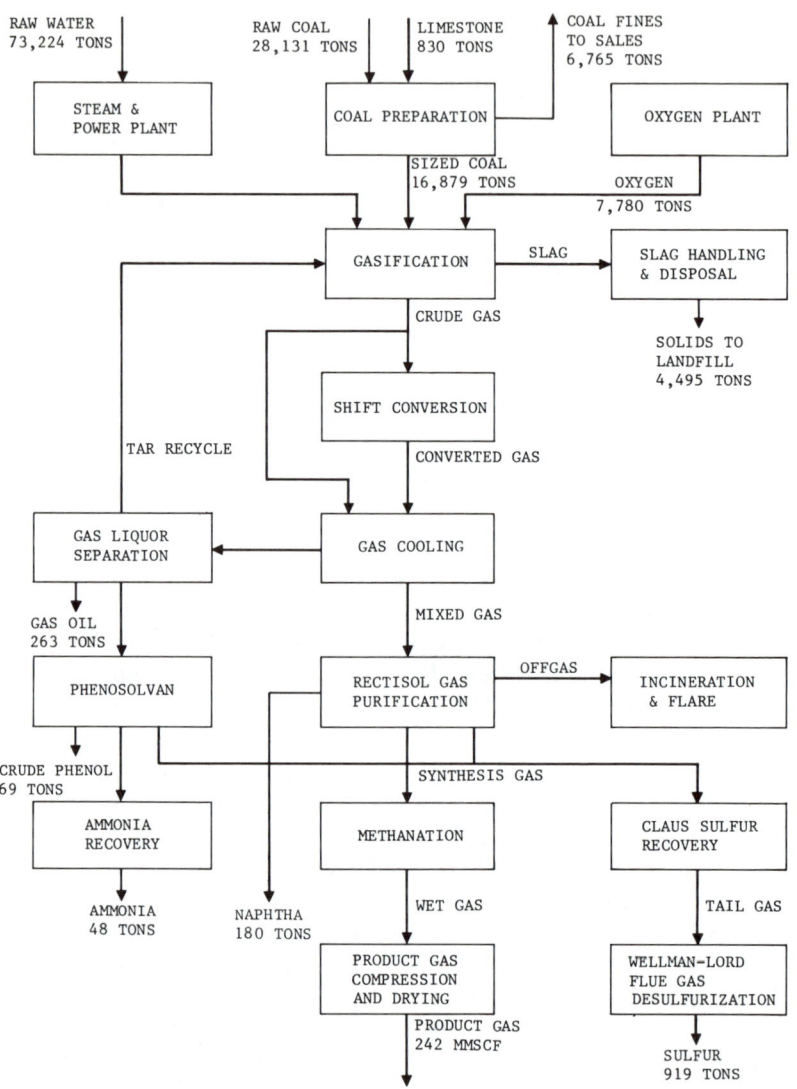

Figure 6-1. Conceptual slagging Lurgi pipeline gas plant (stream-day basis).

Another feature unique to the slagging Lurgi is the way in which the oxygen and steam are injected into the gasifier. In the slagging Lurgi, the gas flow is directed downward, close to the exit, to permit the ashes to leave the gasifier while still molten. In the dry-ash Lurgi, the gases flow upward, well above the exit, to permit the ashes to cool below the clinkering point before they exit the gasifier.

The pilot-plant operation provided several thousand hours of operating experience with the gasifier, purification system and finally the complete system including the Conoco developed methanator. Pipeline quality gas with 960 Btu/scf (36 MJ/normal m^3) heat content was produced and fed into the commercial gas network. The next step in the development program will be to build a demonstration plant using equipment of commercial size. The gasifier will be about 10 ft (3 m) in diameter and have a gas throughput 2.6 times the pilot plant unit. As part of the pilot plant study a large commercial size facility producing 242 × 10^6 scf (6.9 MJ/10^6 normal m^3) of pipeline gas per day was designed. The technical description, economic analysis and environmental discussion are taken from the design study of this conceptual plant.

Design of the Commercial Plant

The slagging Lurgi commercial plant will consist of nine operating gasifiers plus three spares feeding into three trains of purification and methanation equipment. Additional plant facilities will include administration facilities, security, water and power stations, cooling towers, utility connections, product storage tanks, and shipping and receiving facilities. The size of the facility may be determined from Table 6-1 which

Table 6-1. Commercial-Size Gasification Plant Inputs and Products

Material	Amount/Stream-Day	
	English Units	Metric Units
Input		
Coal (Illinois No. 6)	16,879 tons	15,313 metric tons
Flux (limestone)	830 tons	753 metric tons
Steam at 550 psig		
(37 atm) and 750°F (400°C)	5,540 tons	5,026 metric tons
Oxygen (98%)	7,779 tons	7,057 metric tons
Recycle Tar	1,224 tons	1,110 metric tons
Products and By-Products		
SNG, 960 Btu/scf		
(36 MJ/normal m^3)	242 million scf	6.9 million normal m^3
Coal Fines	6,764 tons	6,136 metric tons
Crude Phenols	69 tons	63 metric tons
Naphtha	180 tons	164 metric tons
Tar Oils	262 tons	238 metric tons
Anhydrous Ammonia	48 tons	44 metric tons
Sulfur	919 tons	834 metric tons

136 GASIFICATION

Table 6-2. SNG Properties

	English Units		Metric Units	
Gross Heating Value	960	Btu/scf	36	MJ/normal m^3
Molecular Weight		16.22		
Water Content	7	lb/10^6 scf	0.1	kJ/normal m^3
H$_2$S, Maximum	0.25	gr/100 scf	5.2	mg/normal m^3
Total Sulfur, Maximum	1.0	gr/100 scr	20	mg/normal m^3
Composition (vol %)				
Methane		94.29		
Hydrogen		2.45		
Carbon Monoxide		0.00		
Carbon Dioxide		0.80		
Nitrogen		2.46		
Total		100.00		

defines the daily input and output. The properties of the synthetic natural gas are listed in Table 6-2.

The commercial plant consists of the following sections:

1. coal and flux handling and preparation,
2. air separation,
3. gasification,
4. shift conversion,
5. gas cooling,
6. Rectisol,
7. methanation,
7. product gas compression and drying,
9. sulfur recovery,
10. slag handling and disposal,
11. gas liquor separation,
12. phenol extraction,
13. ammonia recovery,
14. water treatment and steam generation,
15. cooling water system,
16. plant and instrument air system,
17. wastewater treatment,
18. flare and incinerator facilities,
19. tankage and fuel oil system, and
20. shipping and receiving facilities.

Proprietary processes used in the plant are as follows:

- gasification: British Gas Corporation, International Consultancy Service;

- shift conversion, gas cooling, Rectisol, gas liquor separation, phenol extraction, ammonia recovery: Lurgi Kohle and Mineraloeltechnik, GmbH;
- methanation Conoco Methanation Company;
- sulfur recovery: Standard Oil Company (Indiana); and
- flue gas desulfurization: Davy Powergas, Inc.

Operating conditions are, of necessity, more stringent in the slagging Lurgi as higher temperatures are needed to facilitate the slag formation. Table 6-3 shows a comparison of operating conditions among a dry-ash Lurgi gasifier, the gasifier utilized at Westfield, and the design parameters for the demonstration plant unit. From the table it can be seen that both slagging gasifiers can operate at much higher temperatures and pressures, with overall better performance as shown by the increased feed rate and gas production per pound of coal, and the reduced steam to oxygen ratio and residence time. In comparing the pilot plant and demonstration plant gasifier it should be noted that the coal feedstocks are not the same. Data for the pilot plant are based on runs with Ohio No. 9 and Pittsburgh No. 8 coals. The demonstration plant has been designed to operate principally with Illinois No. 6 coal.

Gas outlet temperatures for the two gasifiers differ because of the longer gas-solid contact time encountered in the demonstration gasifier; the coal bed temperature is the same for both as the slag is removed from the units by the same technique. The demonstration gasifier operates at higher pressures to increase its coal processing rate (throughput). The demonstration plant exhibits a somewhat lower overall efficiency than the pilot plant because Illinois No. 6, the coal feedstock, contains lesser amounts of volatiles and fixed carbon than do the coals used in the pilot plant.

In this process, coal sized to 2 by 0.25 in. (5 by 0.6 cm), recycled tar and a flux agent (limestone) are fed, batchwise, through lock hoppers to the gasifier. The function of the limestone is to control the slag viscosity so as to enhance the tapping operation. The tar is added to enhance the process economics, and to help control coal and char dust carryover with the product gases and to dispose of solids which accumulate in the tar. When the loading operation is completed, the pressure in the coal lock is raised to that of the gasifier with recycle lock gas before the coal is dropped into the distributor and stirrer section of the gasifier. On unloading, the coal lock is depressurized by venting the gas into the lock gas holder and made ready for the next batch. The coal mixture is distributed evenly over the entrained coal bed by means of the distributor. The formation of large voids and coal agglomerates is minimized by the use of a rotating stirrer.

Table 6-3. Gasifier Operating Condition

	Slagging Lurgi Demonstration Plant Illinois No. 6	Westfield Pilot Plant Pittsburgh No. 8 and Ohio No. 9	Dry-Ash Lurgi Commercial Plant Pittsburgh No. 8
Coal Feed			
Gas Outlet Temperature [°F (°C)]	331 (166)	770–960 (410–515)	850 (454)
Coal Bed Temperature [°F (°C)]	2600°F (1371)	1800–2500 (982–1371)	980–1370 (526–743)
Gasifier Pressure [psia (atm)]	450 (31)	350 (24)	300 (20)
Coal Residence Time (min)	NA[a]	10–15	60
Steam-to-Oxygen Ratio (mol/mol)	1.27	1.15–1.45	8–10
Oxygen (ton/ton coal)[b]	0.46	0.56	0.70
Steam (ton/ton coal)[b]	0.33	0.41	5.10
Calorific Value of Finished Product [Btu/scf (MJ/normal m³)]	960 (36)		450–570 (17–21)
Cold Gas Efficiency (%)	87.2	83	60–63
Gas Production Rate [scf/lb coal (normal m³/kg)]	33 (198)	34–36 (204–216)	16–30 (96–130)
Coal Feed Rate [lb/hr/ft² (kg/hr/m²)][b]	966 (40–8)	670–870 (28.3–36.8)	100–400 (4.22–16.9)
Raw Gas Heating Value [Btu/scf (MJ/normal m³)]	360 (13)	359 (13)	286 (11)

[a]NA = not applicable.
[b]Moisture and ash free.

MAJOR GASIFICATION PROCESSES 139

Oxygen (98% pure), and superheated steam at 550 psig (37 atm) are injected as one stream through tuyeres located at the bottom of the gasifier.

The coal entering the reactor moves to the bottom of the gasifier in a slowly moving bed. As it travels countercurrent to the rising product gases, volatile matter and the moisture in the coal are driven off and carried away by the hot rising product gases. Gasification of the coal occurs closer to the bottom of the gasifier. The coal first reacts with the oxygen being injected to generate the necessary heat needed for the reaction of the carbon with steam to produce the synthesis gas (CO, H_2).

$$C + H_2O \rightarrow CO + H_2$$

The gasification process is completed by reacting carbon monoxide with steam to produce more hydrogen, and by reacting the carbon still unreacted with the hydrogen just produced.

$$CO + H_2O \rightarrow CO_2 + H_2$$
$$C + 2H_2 \rightarrow CH_4$$

The gasifier is capable of processing all coals, including Western lignite; however, the pilot plant trials were conducted with Pittsburgh No. 8 coal and to a lesser extent with Ohio No. 9 coal. Enough experience was gained from these trials, that operability of the gasifier can be predicted with confidence. The output of the gasifier is not dependent on the type of coal used. Material balance studies of the commercial plant gasifier performance show no change in the gas production rates or the heating value of the gas produced. These results are obtained by changing the feed rates to optimize gas production. Table 6-4 shows the proximate and ultimate analyses of the different coals that could be utilized in the plant as well as the operating conditions when using these coals. From the table it can be seen that Illinois No. 6 coal is of the lowest rank as evidenced by the low carbon content and low heating value. This is reflected on the greater amount of material that must be fed to the gasifier and steam generation units to produce gas and steam at the same rate as the other coals.

Feed rates to the gasifier vary according to the carbon content of the coal used but in accordance to the ratio established at Westfield during the Technical Support Program (see Pilot-Plant Operations). The rate of production of SNG, however, is the same for all coals. About one million

Table 6-4. Alternative Coal Operating Conditions

	Illinois No. 6	Ohio No. 9	Pittsburgh No. 8
Proximate Analysis (wt %)			
Moisture	12.08	2.50	4.58
Ash	13.27	22.50	7.74
Volatile	30.80	35.00	37.37
Fixed Carbon	43.85	40.00	50.31
Total	100.00	100.00	100.00
Ultimate Analysis (DAF Basis)[a] (wt %)			
Carbon	76.55	78.00	84.57
Hydrogen	5.26	5.65	5.75
Oxygen	10.92	8.75	5.11
Nitrogen	1.11	1.25	1.54
Sulfur	5.95	6.30	2.87
Chlorine	0.21	0.05	0.16
Total	100.00	100.00	100.00
Coal Heating Value (DAF Basis) [Btu/lb (MJ/kg)]	13,650 (31)	14,560 (33)	14,087 (32)
Coal Gross Heating Value (as Received) [Btu/lb (MJ/Kg)]	10,190 (23)	10,920 (25)	13,441 (31)
Gasification Plant Operating Condition			
Coal to Gasifiers [lb/hr (kg/hr)]	1,406,566 (638,018)	1,328,500 (602,607)	1,038,965 (471,274)
Fines to Steam Generation [lb/hr (kg/hr)]	374,000 (169,646)	324,700 (147,283)	287,530 (130,423)
Total Coal to Plant [lb/hr (kg/hr)]	2,344,250 (1,063,352)	2,214,167 (1,004,346)	1,731,583 (785,446)
Steam to Gasifiers [lb/hr (kg/hr)]	461,673 (209,414)	407,255 (184,730)	455,430 (206,583)
Oxygen to Gasifiers [10^6 scf/hr (10^4 normal m^3)]	7.707 (21.826)	6.844 (19.382)	7.530 (21.324)
Acid Gas to Sulfur Recovery (mol/hr)	3,533	3,818	1,584
Ammonia Production [lb/hr (kg/hr)]	4,010 (1819)	1,633 (741)	4,846 (2198)
SNG Production Rate [10^6 scf/day (10^6 normal m^3/day)]	241.7 (6.7)	241.7 (6.7)	241.7 (6.7)
SNG GHV [Btu/scf (MJ/normal m^3)]	960.0 (36)	960.0 (36)	960.0 (36)

scf/hr (28,000 normal m³/hr) of gas is produced with a gross heating value of 960 Btu/scf (36 MJ/normal m³).

The gasifier total material balance is shown in Table 6-5. Major inputs for gas production consist of the coal, oxygen and steam streams. Flux is fed to the gasifier to aid in the slag removal operation. Clear tar and dusty recycle tar streams are fed to the gasifier to enhance its performance.

A detailed flow diagram and stream analysis for the gasification section are shown in Figure 6-2 and in Table 6-6.

The raw gas existing in the gasifier is scrubbed, quenched and saturated by injection water from the gas liquor separation section to remove coal dust and heavy tars. The gas has a medium-Btu value.

Table 6-5. Material Balance: Gasification

	Temperature		lb/hr	kg/hr
	°F	°C		
Input			1,475,720	669,400
Sized Coal and Flux	77	25	1,475,720	669,400
Superheated High-Pressure Steam	750	400	461,673	209,414
Oxygen	275	135	648,288	294,063
Fuel Gas	102	39	1,090	494
Carbon Dioxide	158	70	216,761	98,322
Dusty Recycle Tar	160	71	58,320	26,454
Clear Tar	160	71	43,680	19,813
High-Pressure Boiler Feed Water	250	121	250,463	113,610
Boiler Feed Water (Quench Makeup)	250	121	5,000	2,268
Filling Water	158	70	375,000	170,100
Cooling Water Blowdown				
(Quench Makeup)	87	31	300,000	13,608
Injection Water	160	71	737,503	334,531
Total Input			4,303,498	1,952,066
Output				
Total Raw Gas	331	166	2,600,041	1,179,378
Dusty Gas Liquor	356	180	947,862	429,950
High-Pressure Carbon Dioxide				
Lock-Hopper Offgas	32	10	104,582	47,438
Low-Pressure Carbon Dioxide				
Lock-Hopper Offgas	68	20	3,965	1,798
Slag and Water	158	70	497,592	225,707
Slag Quench Drains	226	108	141,000	63,957
Vent Gas	250	121	1,161	526
Jacket Blowdown	457	236	7,295	3,309
Total Output			4,303,498	1,952,066

142 GASIFICATION

Table 6-6. Stream Analysis: Gasification

Stream	mol wt	mol/hr	mol %	lb/hr	kg/hr	Miscellaneous
300—Sized Coal and Flux to Gasifier						
Coal[a]				1,050,000	476,280	
Ash				186,613	84,647	
Flux				69,154	31,368	
Water	18.016			169,953	77,090	
Total Stream				1,475,720	669,386	
Temperature [°F (°C)]						77 (25)
306—Dustry Recycle Tar from Gas/Liquid Separator						
Tar				45,000	20,412	
Dust				13,320	6,042	
Total Stream				58,320	26,453	
Temperature [°F (°C)]						160 (71)
Pressure [psig (atm)]						150 (10)
308—Clear Tar from Gas/Liquid Separator						
Tar				42,000		
Dust				1,680		
Total Stream				43,680		
Temperature [°F (°C)]						160 (71)
Pressure [psig (atm)]						150 (10)
309—Oxygen to Gasifiers						
Oxygen	32.000	19,903.4	98.00	636,909		
Nitrogen	28.014	406.2	2.00	11,379		
Total Dry Gas		20,309.6	100.00	648,288		
Temperature [°F (°C)]						275 (135)
Pressure [psig (atm)]						500 (34)
Dry Gas [10^6 scf/day (10^6 normal m^3/day)]						185 (5.24)
315—Total Raw Gas Production						
Hydrogen	2.016	23,702.5	25.69	47,784	21,674	
Carbon Monoxide	28.011	53,997.8	58.52	1,512,532	686,084	
Carbon Dioxide	44.011	5,944.1	6.44	261,606	118,664	

Methane	16,043	5,620.8	6.09	90,174	40,902
C_nH_m		465.4	0.50	16,687	7,569
Nitrogen	28.014	656.0	0.71	18,378	8,336
Hydrogen Sulfide	34.080	1,779.3	1.93	60,640	27,506
Organic Sulfur		108.3	0.12	6,643	3,013
Total Dry Gas		92,274.2	100.00	2,014,444	913,751
Water	18.016	30,281.0		545,543	247,458
Total Wet Gas		122,555.2		2,559,987	1,161,210
Other Components				40,054	18,168
Total Stream				2,600,041	1,179,378
Temperature [°F (C°)]					331 (166)
Pressure [psig (atm)]					422 (29)
Dry Gas [10^6 scf/day (10^3 normal m³/day)]					840 (24)
Dry Gas GHV [Btu/scf (J/normal cm³/day)]					357 (13)
320 (1001—Slag and Water to Slag Handling					
Water	18.016			269,000	122,018
Slag				228,592	103,689
Total Stream				497,592	225,707
Temperature [°F (°C)]					158 (70)
Pressure [psig (atm)]					12 (0.8)
330 (1100)—Dusty Gas Liquor to Gas Liquor Separator					
Hydrogen	2.016	6.0	5.80	12	5
Carbon Monoxide	28.011	11.7	11.32	327	148
Carbon Dioxide	44.011	41.9	40.52	1,844	836
Methane	16.043	1.4	1.35	22	10
Hydrogen Sulfide	34.080	42.4	41.01	1,445	655
Total Dry Gas		103.4	100.00	3,650	1,656
Water	18.016			828,218	375,679
Total Wet Gas				831,868	377,335
Other Components				115,994	52,615
Total Stream				947,862	429,949
Temperature [°F (°C)]					356 (180)
Pressure [psig (atm)]					250 (17)

[a]Dry, ash-free basis.

144 GASIFICATION

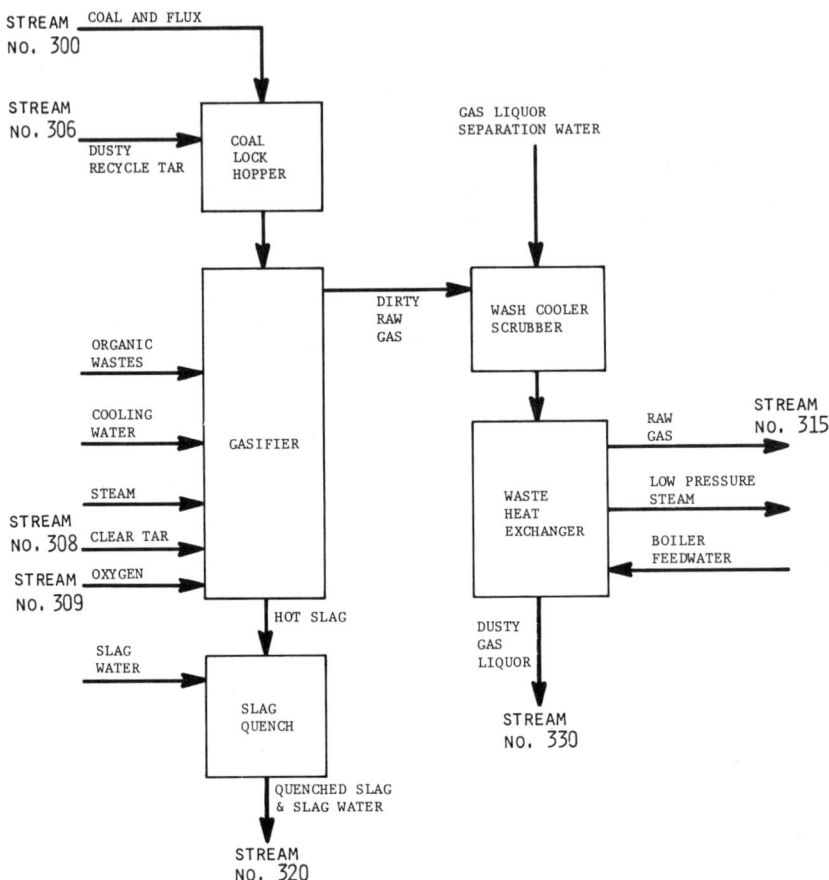

Figure 6-2. Gasification.

For high-Btu gas manufacture, the raw gas needs to be shifted and methanated. In the demonstration plant the raw gas is passed through a waste heat exchanger before it is treated in the shift conversion section. The gas coming to this unit is divided into two streams. The smaller of the two is sent to the gas cooling section for removal of heavy hydrocarbons by condensation. The other stream is sent to the shift conversion unit to adjust the hydrogen–to–carbon monoxide ratio. This stream is divided into three sections and sent into three reactors operating in this section. The stream going to the first reactor is cooled and scrubbed with injection water from the gas liquor separation section. Gas from the first reactor is mixed with clean condensate to control the temperature of the gas before it enters the second reactor. Steam at 550

MAJOR GASIFICATION PROCESSES 145

psig (37 atm) and 750°F (399°C) in excess of the stoichiometric requirements is injected to drive the equilibrium toward production of carbon dioxide and hydrogen.

The operation is carried over a cobalt-molybdenum catalyst fixed-bed reactor. The conversion proceeds according to the reaction:

$$CO + H_2O \rightarrow CO_2 + H_2$$

Due to the nature of the fluids processed through the unit, the catalyst used in the reactors will have to be regenerated periodically. The length of time between regeneration is dependent on the properties of the gas impurities. Catalyst contamination is indicated by increases in pressure drops across the unit and a decrease of catalyst activity. The material balance for the shift conversion section is shown in Table 6-7.

A detailed flow diagram and stream analysis for the shift conversion section is shown in Figure 6-3 and in Table 6-8.

The shifted and bypass gases are cooled in the gas cooling unit to remove the condensables and to utilize the sensible heat in the raw gases. The usable heat removed from the gases is used to superheat high-pressure steam, to produce high- and medium-pressure steam and to heat turbine condensate and boiler feed water. The heat recovery operation is performed in seven stages, with sixth and seventh stages not recovering useful heat. Table 6-9 shows the stages and the function associated with each one.

The outputs from this section consist of a mixed gas stream at 95°F (35°C) going to the Rectisol section as shown in Table 6-10 and

Table 6-7. Material Balance: Shift Conversion

	Temperature			
	°F	°C	lb/hr	kg/hr
Input				
Raw Gas	331	166	2,534,232	1,149,528
Injection Watson	160	71	15,000	6,804
Superheated High-Pressure Steam	750	398	914,894	414,996
Clean Condensate	100	38	3,494	1,585
Total Input			3,467,620	1,572,912
Output				
Converted Gas	878	470	3,382,618	1,534,356
Oil Gas Liquor	248	120	85,002	38,557
Total Output			3,467,620	1,572,912

146 GASIFICATION

Table 6-8. Stream Analysis: Shift Conversion

Stream	mol wt	mol/hr	mol %	lb/hr	kg/hr	Miscellaneous
401—Raw Gas from Gasification						
Hydrogen	2.016	23,102.3	25.69	46,574	21,126	
Carbon Monoxide	28.011	52,631.0	58.52	1,474,247	668,718	
Carbon Dioxide	44.011	6,793.7	6.44	254,985	115,661	
Methane	16.043	5,478.6	6.09	87,892	38,867	
C_nH_m		453.7	0.50	16,265	7,377	
Nitrogen	28.014	639.4	0.71	17,913	8,125	
Hydrogen Sulfide	34.080	1,734.3	1.93	59,105	26,801	
Organic Sulfur		105.6	0.12	6,475	2,937	
Total Dry Gas		89,938.6	100.00	1,963,456	890,623	
Water	18.016	29,514.5		531,734	241,194	
Total Wet Gas		119,453.1		2,495,190	1,131,918	
Other Components				39,042	17,709	
Total Stream				2,534,232	1,149,527	
Temperature [°F (°C)]						331 (166)
Pressure [psig (atm)]						421 (29)
Dry Gas [10⁶ scf/day (10³ normal m³/day)]						819 (23)
Dry Gas GVH [Btu/scf (MJ/normal cm³)]						357 (13)
410—Raw Gas Bypass to Gas Cooling						
Hydrogen	2.016	600.2	25.69	1,210	548	
Carbon Monoxide	28.011	1,366.8	58.52	38,285	17,366	
Carbon Dioxide	44.011	150.4	6.44	6,621	3,003	
Methane	16.043	142.2	6.09	2,282	1,034	
C_nH_m		11.7	0.50	422	191	
Nitrogen	28.014	16.6	0.71	465	211	

MAJOR GASIFICATION PROCESSES 147

Hydrogen Sulfide	34.080	45.0	1.93	1,535	696
Organic Sulfur		2.7	0.12	168	76
Total Dry Gas		2,335.6	100.00	50,988	23,128
Water	18.016	766.5		13,809	6,263
Total Wet Gas		3,102.1		64,797	29,392
Other Components				1,012	459
Total Stream				65,809	29,850
Temperature [°F (°C)]					331 (166)
Pressure [psig (atm)]					421 (29)
Dry Gas [10⁶ scf/day (10³ normal m³/day)]					21 (0.6)
Dry Gas GVH [Btu/scf (MJ/normal cm³)]					356 (13)

411—Converted Gas to Gas Cooling

Hydrogen	2.016	59,076.5	46.81	119,098	54,022
Carbon Monoxide	28.011	16,237.0	12.86	454,815	206,304
Carbon Dioxide	44.011	42,259.1	33.48	1,859,864	843,634
Methane	16.043	5,833.7	4.62	93,590	43,452
C_nH_m		314.1	0.25	11,409	5,175
Nitrogen	28.014	639.4	0.51	17,913	8,125
Hydrogen Sulfide	34.080	1,824.7	1.45	62,186	28,207
Organic Sulfur		21.0	0.02	1,271	576
Total Dry Gas		126,206.3	100.00	2,620,146	1,188,498
Water	18.016	40,284.5		725,766	329,207
Total Wet Gas		166,490.8		3,345,912	1,517,705
Other Components				36,706	16,649
Total Stream				3,382,618	1,534,355
Temperature [°F (°C)]					878 (470)
Pressure [psig (atm)]					379 (26)
Dry Gas [10⁶ scf/day (10³ normal m³/day)]					1,149 (32)
Dry Gas GHV [Btu/scf (J/normal cm³)]					255 (9)

Table 6-8, continued

Stream	mol wt	mol/hr	mol %	lb/hr	kg/hr	Miscellaneous
412—Oily Gas Liquor to Gas/Liquor Separation						
Hydrogen	2.016	0.5		1	0.5	
Carbon Monoxide	28.011	0.9	2.80	24	11	
Carbon Dioxide	44.011	8.9	4.81	391	177	
Methane	16.043	0.1	49.69	2	1	
Hydrogen Sulfide	34.080	7.5	0.67	256	116	
Total Dry Gas		17.9	42.03	674	306	
Water	18.016		100.00	83,703	37,967	
Total Wet Gas				84,377	38,273	
Other Components				625	283	
Total Stream				85,002	39,556	
Temperature [°F (°C)]						248 (120)
Pressure [psig (atm)]						250 (17)

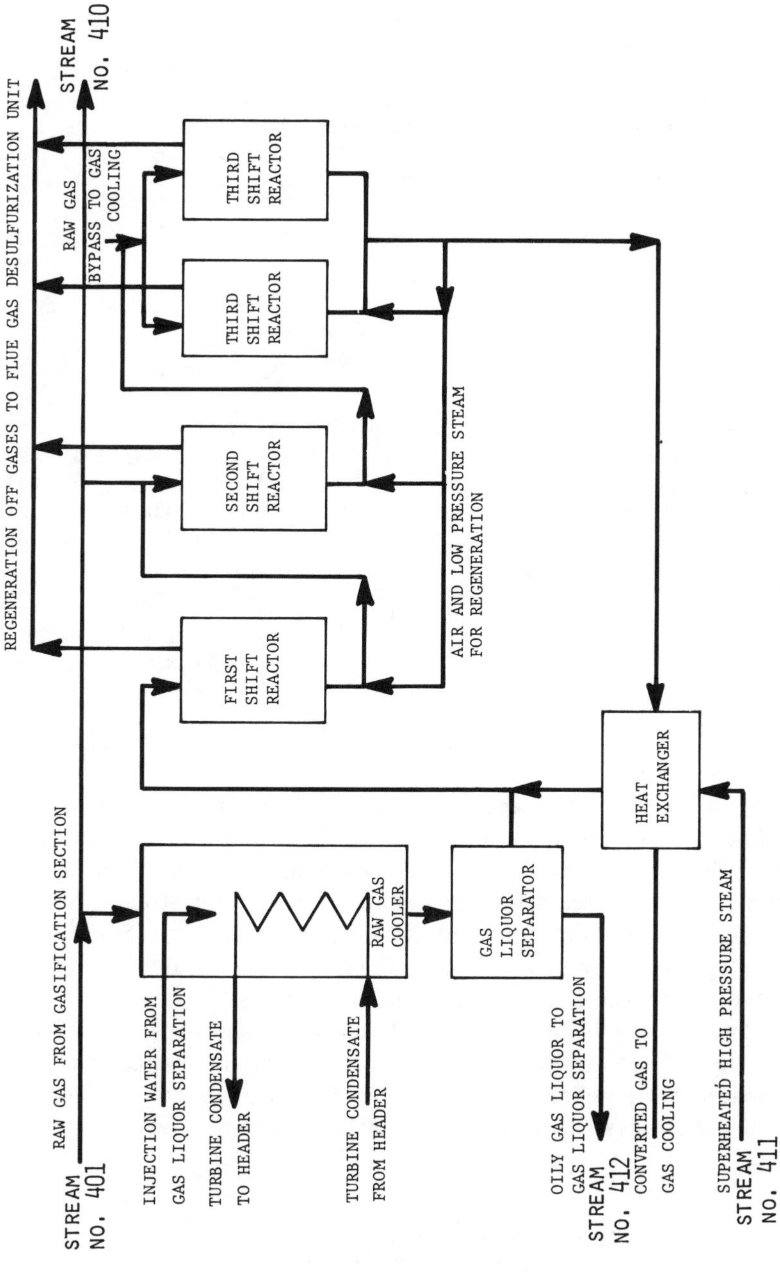

MAJOR GASIFICATION PROCESSES 149

Figure 6-3. Shift conversion.

150 GASIFICATION

Table 6-9. Heat Recovery

1st Stage	550 psig (37 atm) steam superheating
2nd Stage	550 psig (37 atm) steam generation
3rd Stage	110 psig (7 atm) steam generation
4th Stage	Turbine condensate preheat
5th Stage	Demineralizer water preheat
6th Stage	Air cooling
7th Stage	Water cooling

Table 6-10. Material Balance: Gas Cooling

	Temperature			
	°F	°C	lb/hr	kg/hr
Input				
Converted Gas	878	470	3,381,618	1,534,355
Bypassed Raw Gas	331	166	65,809	29,850
Injection Water from Section 1100C	160	71	120,000	54,432
Stripped Gas Liquor from Section 1300C	205	96	120,000	54,432
Total Input			3,688,427	1,673,070
Output				
Mixed Gas	95	35	2,645,826	1,200,146
Oily Gas Liquor	226	107	1,042,601	472,923
Total Output			3,688,427	1,673,070

an oily gas liquor stream at 226°F (107°C) going to the gas liquor recovery section.

The cooled gases are sent to the Rectisol unit for purification and adjustment in the composition of the synthesis gas. Purification of the gas is achieved by removing sulfur compounds, carbon dioxide and naphtha from the stream with methanol. The resulting streams are sent to the phenol extraction unit and the sulfur recovery unit. Some of the carbon dioxide is used to pressurize the coal locks on the gasifier. The methanol is recovered and recycled into the process.

A stream analysis of the Rectisol section is shown in Table 6-11.

The gases from the Rectisol unit are sent to the methanation unit for conversion into pipeline quality gas. The following reactions take place over nickel catalyst, temperature-controlled, fixed-bed reactors:

$$CO + 3H_2 \rightarrow CH_4 + H_2O$$
$$CO_2 + 4H_2 \rightarrow CH_4 + H_2O$$
$$C_nH_m + (2n - m/2)H_2 \rightarrow nCH_4$$

The heat from the above reactions is used to produce saturated 600-psig steam for use in the boilers.

The material balance of the methanation section is shown in Table 6-12.

A steam analysis for the methanation section is shown in Table 6-13.

Prior to releasing the gas for distribution, it is necessary to remove environmentally objectionable sulfur oxides. This operation is accomplished by passing the gas through the sulfur recovery unit. In this unit, the following reaction takes place:

$$2H_2S + SO_2 \rightarrow 3S + 2H_2O$$

Before the above reaction can take place, it is necessary to have the hydrogen sulfide/sulfur dioxide in a 2:1 mol/mol ratio to obtain maximum sulfur recovery. The ratio is achieved by burning some of the hydrogen sulfide with oxygen from the air.

The sulfur recovery unit is fed by streams coming from four sections of the plant: the Rectisol acid gas, phenol extraction acid gas, boiler fuel gas SO_2, and the gas liquor separation expansion gas. Material balance for the sulfur recovery section is shown in Table 6-14.

The liquid sulfur flows by gravity to a central collection unit where the temperature is maintained at 275°F (135°C) to facilitate pumping to a storage area.

The dusty and oily gas liquors from the gasification and shift conversion sections respectively are collected in the gas liquor separation unit, and the dust, tars, oils, phenol and ammonia contained in those gases are removed sequentially.

Pilot-Plant Operations

Pilot-plant operations were carried out at the Westfield Development Centre under the auspices of the Technical Support Program (TSP) from July 1976 to August 1978. During this period, 15 trial runs were conducted with the following results:

- The necessary data required to design the demonstration plant were obtained.
- Operability of the gasifier was demonstrated with caking Eastern U.S. coals and with up to 23% coal fines in the feed.

152   GASIFICATION

Table 6-11. Stream Analysis: Rectisol

Stream	mol wt	mol/hr	mol %	lb/hr	kg/hr	Miscellaneous
601—Mixed Gas from Cooling						
Hydrogen	2.016	59,662.7	46.80	120,280	54,559	
Carbon Monoxide	28.011	17,599.2	13.81	492,970	223,611	
Carbon Dioxide	44.011	41,463.2	32.52	1,824,838	827,746	
Methane	16.043	5,972.3	4.69	95,814	43,461	
C_nH_m		326.6	0.26	11,831	5,366	
Nitrogen	28.014	656.0	0.51	18,378	8,336	
Hydrogen Sulfide	34.080	1,765.6	1.39	60,173	27,294	
Organic Sulfur		23.8	0.02	1,439	652	
Total Dry Gas		127,469.4	100.00	2,625,723	1,191,028	
Water	18.016	286.1		5,154	2,337	
Total Wet Gas		127,755.5		2,630,877	1,193,366	
Ammonia	17.031			1	0.5	
Naphtha				14,948	6,780	
Total Stream				2,645,826	1,200,147	
Temperature [°F (°C)]						95 (35)
Pressure [psig (atm)]						355 (24)
Dry Gas [10⁶ scf/day (10³ normal m³/day)]						1,160 (32)
Dry Gas GHV [Btu/scf (J/normal cm³)]						264 (10)
602—Synthesis Gas to Methanation						
Hydrogen	2.016	59,594.3	69.77	120,142	54,496	
Carbon Monoxide	28.011	17,525.9	20.52	490,918	222,680	
Carbon Dioxide	44.011	1,771.6	2.07	77,965	35,364	
Methane	16.043	5,744.4	6.72	92,158	51,802	
C_nH_m		127.3	0.15	3,825	1,735	
Nitrogen	28.014	654.8	0.77	18,345	8,321	

MAJOR GASIFICATION PROCESSES

Total Stream		85,418.3	100.00	803,353	364,400	
Temperature [°F (°C)]						73 (23)
Pressure [psig (atm)]						280 (19)
Dry Gas [10^6 scf/day (10^3 normal m^3/day)]						778 (22)
Dry Gas GHV [Btu/scf (J/normal cm^3)]						363 (13)
604—Acid (Claus) Gas to Sulfur Recovery						
Carbon Dioxide	44.011	1,701.0	48.15	74,864	33,958	
C_nH_m		49.0	1.38	2,666	1,209	
Hydrogen Sulfide	34.080	1,765.6	49.98	60,173	27,924	
Organic Sulfur		17.4	0.49	1,044	473	
Total Stream		3,533.0	100.00	138,747	62,395	
Temperature [°F (°C)]						68 (20)
Pressure [psig (atm)]						10 (0.7)
Dry Gas [10^3 normal m^3/day]						32 (0.9)
Dry Gas GHV [Btu/scf (J/normal cm^3)]						366 (13)
605—Gas Liquor to Gas Liquor Separation						
Water	18.016			3,865	1,753	
Total Stream				3,865	1,753	
Temperature [°F (°C)]						60 (16)
Pressure [psig (atm)]						230 (16)
607—Naphtha Product to Tankage						
Organic Sulfur		0.5	100.00	39	17	
Total Dry Gas		0.5	100.00	39	17	
Ammonia	17.031			1	0.5	
Naphtha				14,948	5,780	
Total Stream				14,988	5,798	
Temperature [°F (°C)]						100 (38)
609—Carbon Dioxide to Gasification						
Hydrogen	2.016	8.9	0.18	18	3	
Carbon Monoxide	28.011	9.4	0.19	264	120	

Table 6-11, continued

Stream	mol wt	mol/hr	mol %	lb/hr	kg/hr	Miscellaneous
Carbon Dioxide	44.011	4,891.3	98.43	215,270	97,646	
Methane	16.043	29.4	0.59	471	213	
C_nH_m		19.3	0.59	687	312	
Nitrogen	28.014	0.2	0.00	5	2	
Organic Sulfur		0.8	0.02	46	21	
Total Stream		4,959.3	100.00	216,761	98,322	
Temperature [°F (°C)]						158 (70)
Pressure [psig (atm)]						500 (34)
Dry Gas [10^6 scf/day (10^3 normal m^3/day)]						45 (1.3)
Dry Gas GHV [Btu/scf (J/normal cm^3)]						16 (0.6)
610—Carbon Dioxide Offgas to Incineration						
Hydrogen	2.016	59.5	0.18	120	54	
Carbon Monoxide	28.011	63.8	0.19	1,788	811	
Carbon Dioxide	44.011	33,099.4	98.63	1,456,739	60,777	
Methane	16.043	198.5	0.59	3,185	1,444	
C_nH_m		130.9	0.39	4,651	2,109	
Nitrogen		1.1	0.00	30	13	
Organic Sulfur		5.2	0.02	310	140	
Total Stream		33,558.4	100.00	1,466,823	665,351	
Temperature [°F (°C)]						68 (20)
Pressure [psig (atm)]						1.2 (0.08)
Dry Gas [10^6 scf/day (10^3 normal m^3/day)]						306 (9)
Dry Gas GHV [Btu/scf (J/normal cm^3)]						18 (0.6)

Table 6-12. Material Balance: Methanation

	Temp		lb/hr	kg/hr
	°F	°C		
Input				
Synthesis Gas	73	23	803,353	364,440
Total Input			803,353	364,440
Output				
Methanation Product	230	110	481,102	218,227
Methanation Condensate	247	119	322,251	146,173
Total Output			803,353	364,440

The goals of the program were:

1. to identify and resolve any problems associated with the operation of the slagging Lurgi gasifier;
2. to provide enough data to permit the design of the demonstration plant with a high degree of confidence;
3. to demonstrate the feasibility of the gasification of moderately caking high-sulfur Eastern U.S. coals; and
4. to obtain stream samples to facilitate an accurate environmental assessment of the process.

To accomplish the program goals, a dry-ash Lurgi gasifier was modified to operate in the slagging mode. Other equipment used in the program included three coal-fired steam boilers, an oxygen plant, a carbon monoxide shift conversion unit, a gas drying plant a sulfur recovery plant, a gas liquor separator plant, a Benzole absorber, and a Rectisol/methanator unit. The Rectisol/methanator unit was originally designed to operate with gas feed from a dry-ash Lurgi gasifier. Its capacity was 2.6×10^3 scf (74 normal m^3) of high-Btu gas from 10.2×10^3 scf (290 normal m^3) of synthesis gas. The unit was successfully used also with the slagging gasifier.

The program looked into the following variables:

1. coal throughput rate as determined by level of oxygen input rate to the gasifier;
2. the level of steam input relative to the level of oxygen input (steam/oxygen ratio);
3. the choice of flux as to type and the addition rate required;
4. feed of recycled solids-laden tars to the top of the gasifier;

Table 6-13. Stream Analysis: Methanation

Stream	mol wt	mol/hr	mol %	lb/hr	Miscellaneous
700—Synthesis Gas from Rectisol					
Hydrogen	2.016	59,594.3	69.77	120,142	
Carbon Monoxide	28.011	17,525.9	20.52	490,918	
Carbon Dioxide	44.011	1,771.6	2.07	77,965	
Methane	16.043	5,744.4	6.72	92,158	
C_nH_m		127.3	0.15	3,825	
Nitrogen	28.014	654.8	0.77	18,345	
Total Stream		85,418.3	100.00	803,353	
Temperature (°F)					73
Pressure (psig)					280
Dry Gas (10^6 scf/day)					778
Dry Gas GHV (Btu/scf)					363
701—Methanation Product to Product Gas Compression and Drying Unit					
Hydrogen	2.016	651.9	2.45	1,314	
Carbon Monoxide	28.011	0.6	0.00	12	
Carbon Dioxide	44.011	211.8	0.80	9,321	
Methane	16.043	25,084.2	94.29	402,426	
Nitrogen	28.014	654.8	2.46	18,345	
Total Dry Gas		26,603.4	100.00	431,418	
Water		2,757.8		49,684	
Total Stream		29,361.2		481,102	
Temperature (°F)					203
Pressure (psig)					208
Dry Gas (10^6 scf/day)					242
Dry Gas GHV (Btu/scf)					960

MAJOR GASIFICATION PROCESSES 157

5. the size consistency of the feed coal, especially with respect to the content of materials less than 0.25 in. (0.6 cm);
6. rotational speed of coal distributor and stirrer; and
7. critical gasifier component operating life.

Results of the TSP trials are summarized below.

The coal throughput rate is a linear function of the oxygen load, with the coal feed rate and oxygen load decreasing with the addition of recycled solids to the feed. The linear dependency and recycled solids effect on the gasifier are shown in Figure 6-4.

The only effect associated with ranging the steam/oxygen ratio from 1.35 to 1.15 mol/mol was that of changing hearth temperatures and consequently varying slag viscosities.

The effect of flux material on slag viscosity was studied throughout the program. Blast furnace slag was used in most of the runs. Limestone was used in run number 4 only. The poor results obtained in this and in runs 6 and 9 are considered to be the result of insufficient flux in the feed.

The two fluxes differ in effect on the slag. Limestone reacts with the coal ash to form a low-viscosity slag rich in calcium and low in silica. The amount of flux required to obtain the appropriate slag viscosity is reduced but the slag viscosity sensitivity to variations in the rate at which limestone is added is increased.

The blast furnace slag melts at the gasifier operating temperatures to form a free-flowing liquid. Since it is deficient in calcium and magnesium, more flux is required to obtain the same results as with lime-

Table 6-14. Material Balance: Sulfur Recovery

	Temp			
	°F	°C	lb/hr	kg/hr
Input				
Acid (Claus) Gas	68	20	138,747	62,935
Acid Gas	95	35	7,654	3,471
Air to Furnaces	77	25	72,257	32,775
Sulfur Dioxide	115	46	34,867	15,816
Expansion Gas	95	35	43,007	19,508
Total Input			296,532	134,507
Output				
Claus Offgas	320	160	219,966	99,776
Sulfur Product	285	141	76,566	34,730
Total Output			296,532	134,507

158 GASIFICATION

stone, but the slag viscosity is not as sensitive to small changes in the rate of addition.

Results from the TSP trials show that a silica ratio less than 55 results in good slag tapping conditions. These results are summarized in Table 6-15 and Figure 6-5. The silica ratio is defined as the ratio of silica to

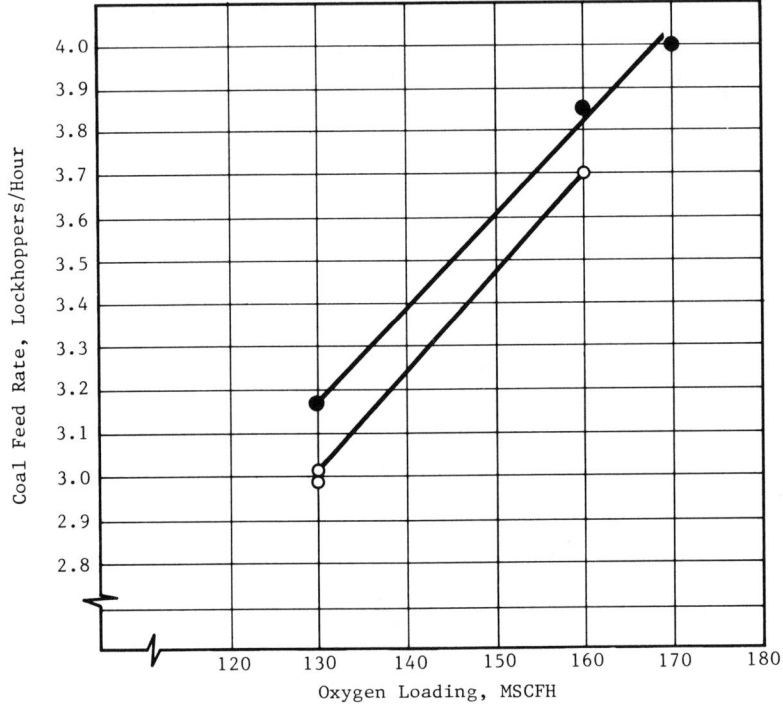

● With Feed of Solids-Laden Tar
○ Without Feed of Solids-Laden Tar
NOTE: All Data are for 1.30 Steam/Oxygen

Figure 6-4. Coal feed rate vs oxygen loading with and without feed or recycle solids-laden tar.

the sum of the weights of all the components present in the slag, multiplied by 100.

$$\text{silica ratio} = \frac{SiO_2}{SiO_2 + CaO + MgO + FeO_3} \times 100$$

The more silica present, the higher the numerical value of the viscosity.

Table 6-15. Slag Tapping Performance for Selected Runs

Run	1	1	2	4	6	9	13	14	15
Feedstock	Frances coal	Frances coal	Ohio No. 9 coal	Ohio No. 9 coal	Ohio No. 9 coal	B.F. coke	Pittsburgh No. 8 coal	Ohio No. 9 coal	Pittsburgh No. 8 coal
Flux	None	B.F. slag	B.F. slag	Limestone	B.F. slag	B.F. slag	B.F. slag	B.F. slag	B.F. slag
Stream/O_2 Ratio	1.35	1.35	1.15	1.25	1.25	1.35	1.30	1.25	1.30
Slag Composition (wt %)									
CaO	14.4	30.5	17.0	13.2	15.0	16.8	26.5	20.4	26.5
MgO	7.4	11.6	6.1	1.3	4.9	5.8	7.8	5.6	7.2
SiO_2	33.0	37.1	43.4	40.7	38.6	46.7	40.1	43.0	40.7
Al_2O_3	22.5	14.8	19.5	18.2	17.3	23.0	18.0	19.0	17.8
Fe_2O_3	14.5	4.7	12.4	16.3	10.1	4.1	5.7	9.7	5.4
Silica Ratio	48	44	55	57	56	64	50	53	50
Hearth Geometry	Normal					Deep			
Slag Tapping Performance	Good	Excellent	Good	Poor	Fair to poor →	Poor	Very good	Fair to good	Very good ↑

160 GASIFICATION

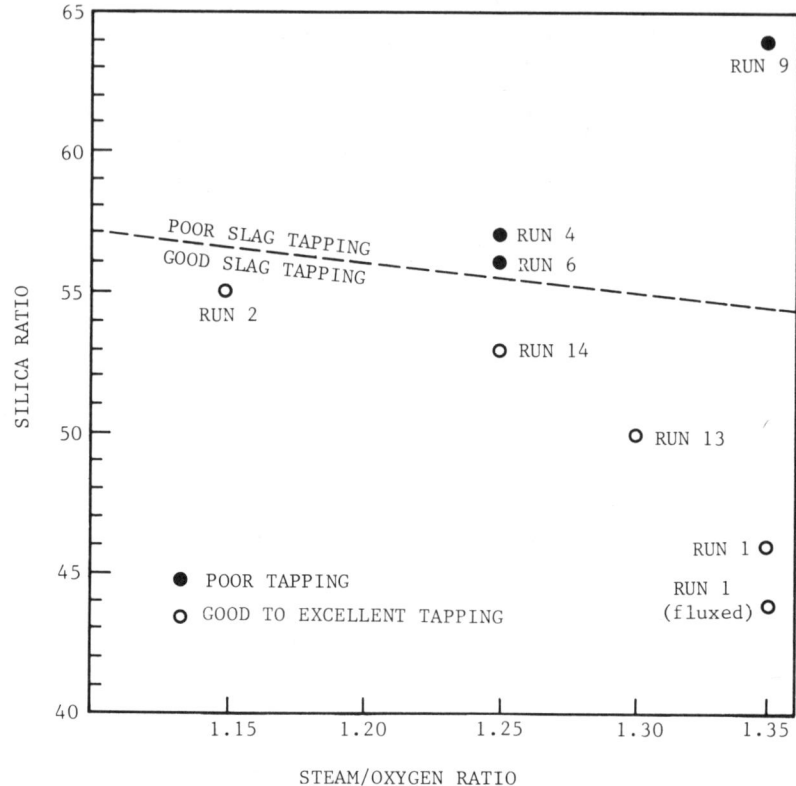

Figure 6-5. Effect of slag composition and steam/oxygen ratio on slag tapping.

The effect of feeding coal fines to the gasifier was found to be insignificant. Little or no performance variation was observed in the gasifier output. In addition, it was found that the level of recycled solid reaches an equilibrium plateau if solids-laden tars are fed to the gasifier, and that the solids content is a function of the amount of coal fines fed and the feed rate of the solids-laden tars.

During the TSP trials, the gasifier was operated with up to 23 percent fines with little or no deviations in its operating parameters. Tables 6-16 and 6-17 show the effect of recycled solids on the composition of the gasifier's output gas. The data were taken from one of the late runs (run 13) using untreated Pittsburgh No. 8 coal.

The torque necessary to operate the distributor/stirrer was found to be a function of the caking characteristics of the feed coals. Increasing

Table 6-16. Effect of Tar Recycle on Product Gas Composition During TSP Run 13: Flare Gas Analysis During Periods of Tar Recycle to the Gasifier[a]

| | 6/19/78 | 6/20/78 | | | | 6/21/78 | | Mean | Standard | 75% Mean Confidence |
	10:40 PM	12:30 AM	4:45 AM	6:40 AM	9:00 AM	4:40 AM	7:30 AM	10:30 AM	Value	Deviation	Limits
CH_4	7.85	6.80	6.57	7.40	7.54	7.05	7.74	6.74	7.21	0.487	±0.22
CO_2	3.11	3.19	3.08	3.50	3.55	3.65	3.76	4.32	3.52	0.412	±0.18
C_2H_4	0.10	0.15	0.19	0.19	0.19	0.27	0.20	0.20	0.19	0.048	±0.02
C_2H_6	0.85	0.47	0.50	0.45	0.46	0.77	0.47	0.49	0.56	0.158	±0.07
H_2S	0.47	0.43	0.53	0.51	0.51	0.55	0.59	0.53	0.52	0.049	±0.02
H_2	27.95	28.76	28.33	28.46	29.54	28.32	28.55	28.82	28.59	0.471	±0.21
O_2	Nil	Nil	Nil	Nil	Nil	Nil	Nil	Nil	Nil		
Ar	0.71	0.68	0.99	0.93	0.92	0.84	0.83	0.92	0.85	0.110	±0.05
N_2	3.39	2.70	3.43	3.70	3.49	3.66	3.68	3.29	3.42	0.326	±0.15
CO	52.92	53.74	54.13	53.33	52.40	52.47	52.52	52.67	53.02	0.645	±0.29
Recovery	97.35	96.92	97.75	98.47	98.60	97.58	98.34	97.98			

[a] 130,000 scf/hr oxygen and 1.30 steam/oxygen.

162 GASIFICATION

Table 6-17. Effect of Tar Recycle on Product Gas Composition During TSP Run 13: Flare Gas Analysis During Periods of No Tar Recycle to the Gasifier[a]

	6/20/78		6/21/78			Mean Value	Standard Deviation	75% Mean Confidence Limits
	1:10 PM	4:30 PM	10:40 PM	12:40 AM	3:10 PM	2:00/4:00 PM		
CH_4	7.04	6.82	7.72	7.27	7.04	6.73	7.10	0.357	±0.10
CO_2	3.64	3.71	3.89	3.52	3.70	3.78	3.71	0.125	±0.07
C_2H_4	0.33	0.13	0.20	0.19	0.29	0.14	0.21	0.081	±0.04
C_2H_6	0.58	0.53	0.53	0.46	1.25	0.46	0.64	0.305	±0.16
H_2S	0.53	0.50	0.51	0.67	0.57	0.53	0.55	0.063	±0.03
H_2	26.76	29.45	28.34	28.88	27.54	28.85	28.30	0.991	±0.53
O_2	Nil	Nil	Nil	Nil	Nil	Nil	Nil		
Ar	0.90	0.90	0.83	0.93	0.88	0.82	0.88	0.043	±0.02
N_2	3.00	2.56	3.18	2.83	2.73	3.77	3.01	0.429	±0.23
CO	54.50	53.38	52.25	53.79	54.48	52.76	53.53	0.913	±0.49
Recovery	97.28	97.98	97.45	98.54	98.48	97.84			

[a]130,000 scf/hr oxygen and 1.30 steam/oxygen.

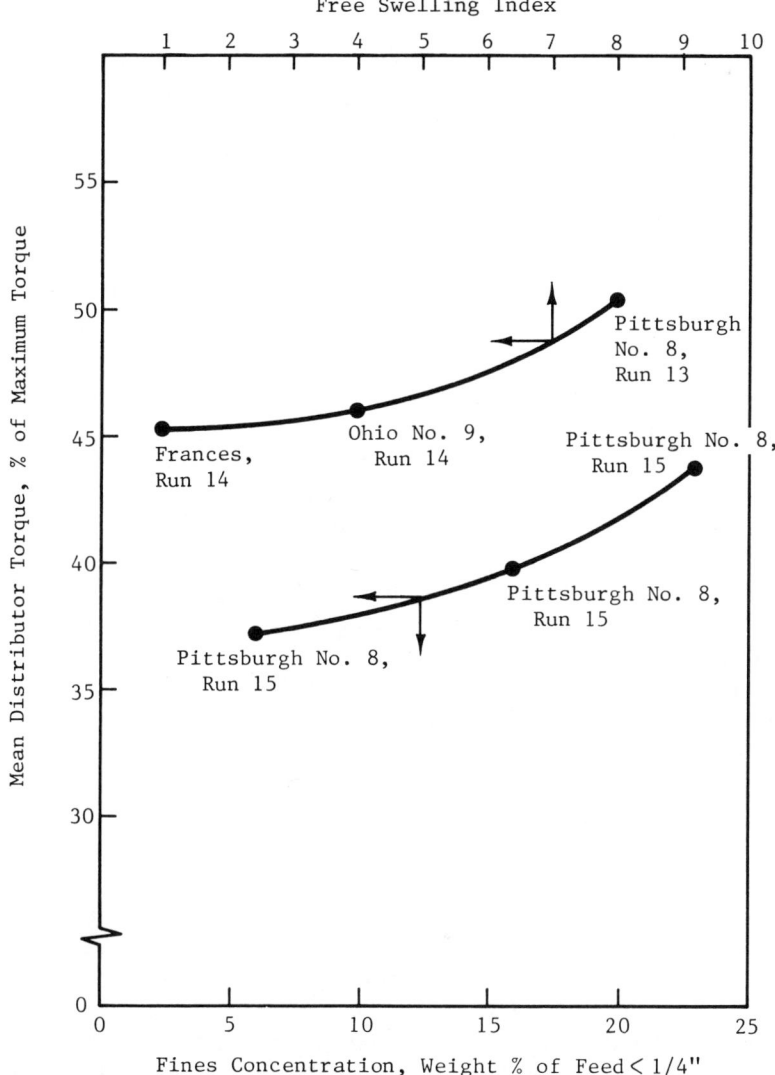

Figure 6-6. Effect of coal swelling index and feed fines concentration on distributor torque.

the concentration of fines in the feed increased the torque required to rotate the distributor assembly. These trends are shown in Figure 6-6.

The gasifier and its components were operated long enough to demonstrate the feasibility of the equipment and to obtain the data necessary to design the demonstration plant. Confidence in their performance

164 GASIFICATION

Table 6-18. Performance of Internal Components of the Pilot-Plant Gasifier

Components	Number of Failures	Demonstrated Hours of Service	Still Usable
Slag Tap	1	670	Yes
Tuyeres	0	1465	Yes
Burner	0	825	Yes
Stirrer	0	670	Yes
Shaft Refactory	0	670	Yes
Hearth Refactory	0	524	Yes

arises from their operation at Westfield and from the experience of the steel industry. The demonstrated hours of operation and the conditions of the components at the end of the TSP are shown in Table 6-18.

The first run of the Technical Support Program was conducted on August 17, 1977. The purpose of this run was to commission the gasifier and its system, both of which had undergone extensive modifications and considerable overhaul since they were last tested in a privately funded program. Major changes to the gasifier include the installation of a new stirrer/distributor system, modifications to the instrumentation to allow automatic control of slag tapping from the main panel board, and installation of a system to feed flux into the gasifier. In this run, the gasifier was tested for operability with Frances coal. Gasifier behavior of lower than normal oxygen load 160,000 scf/hr (4.5×10^3 normal m^3) and at higher hearth temperatures was also investigated.

Operating conditions at which to change from Frances coal to Ohio No. 9 coal were determined to be 130,000 scf/hr (3.7×10^3 normal m^3/hr) of oxygen; steam-to-oxygen ratio, 1.15; and 350 psig (24 atm) gasifier pressure.

During this run, the gasifier operated for 97 consecutive hours. Post-run inspection of the gasifier internals showed no significant damage to internals. The flux feeding mechanism and the stirrer distributor system functioned free of problems.

Massive caking below the stirrer was experienced in five consecutive runs (runs 2, 3, 4, 5 and 6) and in runs 8 and 12. During runs 2 through 6, the intention was to find a solution to the caking problem through manipulation of the gasifier variables. Some success was obtained by lowering the oxygen load. The technique consisted of reducing the oxygen input rate (hence, the coal throughput rate) so as to increase the residence time of the coal in the region affected by the stirrer blades. On run 6, it was concluded that a longer stirrer was needed to provide a long-term solution to the problems presented by the caking coals.

MAJOR GASIFICATION PROCESSES   165

The stirrer was installed and tested during run 8 of the TSP. During run 7 of the program, a new drive had been installed. The purpose of the new drive was to operate the stirrer at more suitable speeds.

Run 8 was prematurely ended due to critical fouling of the quench chamber. This resulted in caking conditions below the stirrer.

Caking conditions were encountered again during run 12; however, at this time, the gasifier was processing Ohio No. 9 coal while the operating conditions were those for processing Pittsburgh No. 8 coal.

Slag deposits were found in the slag quench chamber in several runs (runs 2, 4, 5, 7 and 8). The problem was severe enough so as to cause the gasifier to be shut down every time the condition was present. These difficulties were overcome by making changes in the quench chamber, lowering the hearth, and by installing an oxygen preheater capable of increasing the steam/oxygen stream temperature by 50°F (10°C).

Problems with the slag tap were encountered in runs 5, 6 and 9. In all instances, the difficulties forced the early shutdown of the operation. From these experiences it was concluded that further modifications to the proprietary slag tap equipment were needed, and that a silica ratio of greater than 55% would result in satisfactory performance of the slag tapping operation.

Operation of the stirrer was faulty during four of the runs (runs 3, 8, 13 and 15). In one instance (run 3), the condition forced the shutdown of the gasifier. All four incidents occurred when changing from non-caking to caking coals. The difficulties are due to improper gasifier operating conditions which called for unusual torque requirements.

Two other problems were experienced during the TSP runs. During run 10, coke sized 2 in. × 0 (2.5 cm × 0) was charged to the gasifier. This feedstock resulted in erratic gasifier behavior with respect to offtake temperature, bed pressure drops and carbon dioxide levels in the synthesis gas. Also, the large amount of fines and subsequent dust carryover resulted in plugging of downstream equipment. This coal feed also caused severe plugging of the tuyeres so that gasifier recovery was not possible. In subsequent runs, the condition was averted by not using this type of feed.

A summary of the problems experienced during the TSP is shown in Figure 6-7.

Operating data concerning the gasifier performance was obtained for Pittsburgh No. 8 and Ohio No. 9 coals from runs 13 and 14, respectively. Since the plant was not fully equipped with measuring instruments, data for the fuel (coke or coal) feed rate, the gas liquor yield, and the tar and oil yield are less accurate than data for other streams.

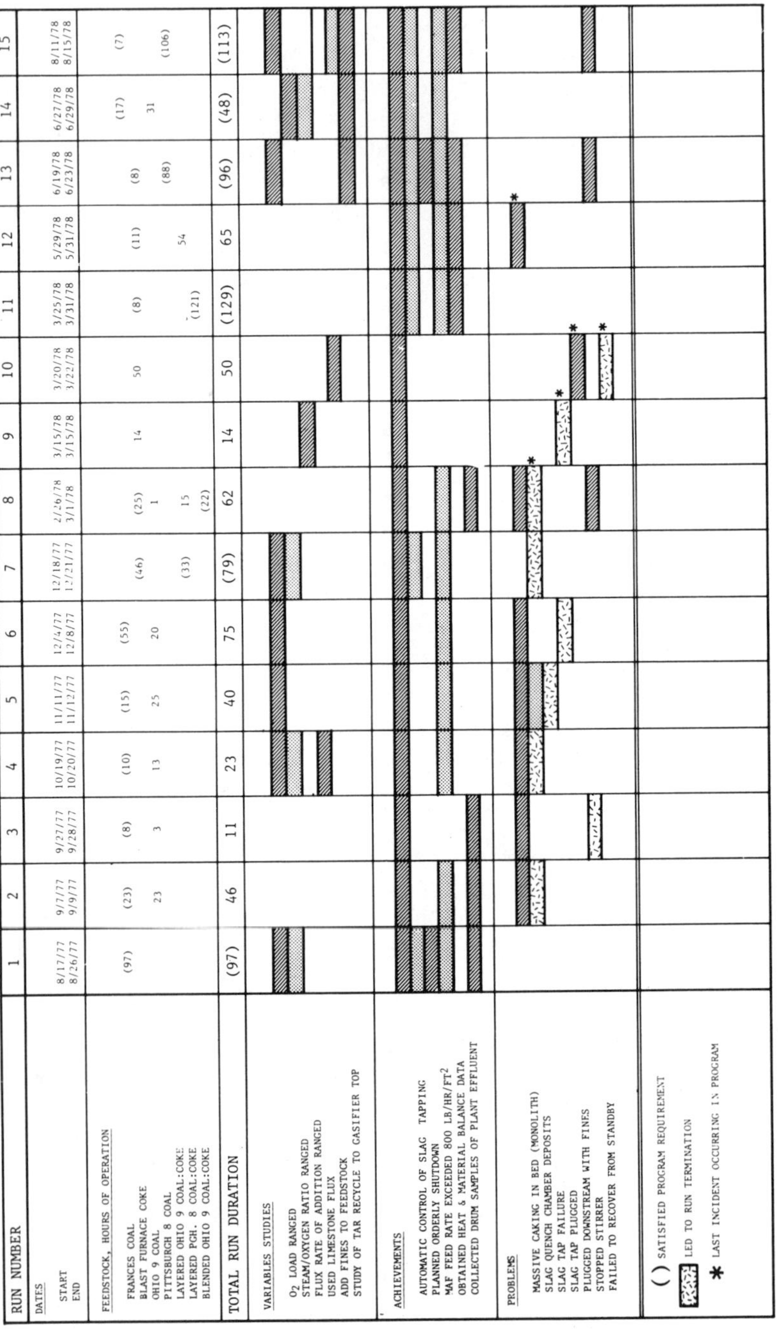

166 GASIFICATION

Figure 6-7. Summary of problems during technical support program.

MAJOR GASIFICATION PROCESSES   167

The fuel feed rate was calculated from the number of lock-hoppers per hour, the volume of the lockhopper, and the bulk density of the feed, with corrections being made for the amount of flux present in the feed. A check on these figures was provided by the weight of the coal transported from the storage pile to the charge hopper.

A combined representative sample of the gas, tar, oil, gas liquor, naphtha (condensables) and dust (powdery char and coal) was taken from a side stream and analyzed. The condensables are condensed at ambient temperature, separated, and weighed.

A summary of the rationalized material and heat balance is shown in Tables 6-19 and 6-20. Both tables reflect operating conditions using

Table 6-19. Summary of Rationalized Material Balances
(Basis: One Ton of Moisture- and Ash-Free Coal or Coke Feed)

Run Number	TSP Run 13	TSP Run 14
Feedstock	Pittsburgh No. 8 coal	Ohio No. 9 coal
Flux	Blast furnace slag	Blast furnace slag
Throughput (lb)		
Feed/hr/ft^2	868	670
Offtake Temperature (°F)	958	770
Input (lb)		
MAF Feedstock	2000	2000
Ash and Flux	551	1009
Oxygen and Air		
O_2	1123	1128
N_2	187	190
Steam	851	813
Fuel Gas	8	10
Total Input	4720	5150
Output (lb)		
Dry Gas		
H_2	108	102
CO	2872	2664
CO_2	319	412
CH_4	214	206
CnHm	38	38
Other compounds	261	317
Subtotal	3812	3739
Gas Liquor Less Input Moisture	191	199
Net Tar and Oil	126	137
Naphtha	32	41
Dust	5	7
Slag	554	1027
Total Output	4720	5150
Input - Output (lb)	0	0

Table 6-20. Summary of Rationalized Heat Balances
(Basis: Coal Higher Heating Value = 100.00)

Run Number Feedstock	TSP Run 13 Pittsburgh No. 8 coal	TSP Run 14 Ohio No. 9 coal
Flux	Blast furnace slag	Blast furnace slag
Throughput (lb)[a]		
Feed/hr/ft^2	868	670
Offtake Temperature (°F)	958	770
Input, Higher Heating Value (%)		
MAF Coal	100.00	100.00
Fuel Gas	.61	0.80
Total Inputs	100.61	100.80
Outputs, Higher Heating Value (%)		
Dry Gas		
H_2	22.41	21.17
CO	42.24	39.53
CH_4	17.24	16.74
CnHm	2.82	2.89
Subtotal	84.71	90.33
Other Gas Compounds	1.28	2.66
Net Tar & Oil	6.92	7.82
Naphtha	1.96	2.50
Dust	0.08	0.17
Slag	.16	0.44
Total Output	95.11	93.92
Sensible and Latent Heat (Output - Input)	3.73	4.45
Heat Loss	1.77	2.43
Total Output Heat	100.61	100.80
Input Heat - Output Heat	0	0

[a] Moisture- and ash-free.

Pittsburgh No. 8 and Ohio No. 9 coals during runs 13 and 14, respectively. The two runs represent the best conditions achieved at Westfield. Blast furnace slag was utilized as flux during both runs.

Technical Evaluation

The slagging Lurgi gasification process is a modification of the older, commercially proven Lurgi dry-ash process that has been in worldwide operation since the 1940s. Developmental work was initiated by the Lurgi Mineraloeltechnik GmbH on a small pilot plant shortly after the end of World War II. The plant was located at the Holten Works of Ruhrchemie AG, Federal Republic of Germany. The unit was sold to the British Gas Corporation (then the Gas Council) which further developed the technology at its Midlands Research Station at Solihull, England, on an experimental basis in the late 1950s and, on a 100-ton/day pilot plant unit in the early 1960s. These trials demonstrated the feasibility of the slagging concept. The project was shelved following discoveries of offshore natural gas reserves.

The gasification program was restarted by the British Gas Corporation in 1974 with sponsorship from Conoco and a consortium of 15 U.S. oil and gas firms. The $10 million program was carried out over a 3-year period and was conducted with the active collaboration of Lurgi Mineraloeltechnik of Frankfurt-am-Main, Federal Republic of Germany.

In late 1977 the U.S. Department of Energy (DOE) awarded a contract to Conoco Oil Company for the design, construction and operation of a commercial-scale plant. Conoco is presently awaiting for DOE to decide on whether to proceed with the construction of the plant.

The design of the plant is based on data obtained from trials of the pilot plant at the Westfield Development Centre located near Cardenden, Scotland. These trials demonstrated that the gasifier could operate with caking coal such as the Pittsburgh No. 8 coal. Operability of the gasifier was demonstrated also with Ohio No. 9 coal, although experience with this coal is somewhat less detailed than with the less caking Pittsburgh No. 8 coal.

Most of the component equipment in the pilot plant has been in commercial operation for a number of years, and there is a wealth of information about performance, and confidence on operability in a commerical size gasification plant. For some components, however, experience is lacking, and there are uncertainties associated with their performance at the scale required for commercial operation. The components and their associated degree of risk is summarized in Table 6-21. The components associated with the process are appraised below.

Materials Preparation

This section relies on well proven technology which has been in use for many years and found to be reliable in a wide variety of operating conditions. For purposes of the demonstration plant, the unit is required to process 17,700 ton/day of coal, which is well within the capabilities of currently operating coal handling equipment.

170 GASIFICATION

Table 6-21. Component Scale-up and Risk Analysis

Component	Scale-Up Factor	Risk
Material Preparation	None	Low
Air Separation	1.3	Low
Gasification	2.8	High
Shift Conversion	10.0[a]	Moderate
Gas Cooling	None	Low
Rectisol	2.5	Low
Methanation	4.0	Moderate
Drying and Compression	None	Low
Sulfur Recovery	None	Low
Slag Handling	None	Low
Gas Liquor Separation	1.3	Low
Phenol Extraction	None	Low
Ammonia Recovery	None	Low
Water Treatment and Steam Generation	1.5	Moderate
Cooling Water System	None	Low
Plant and Instrument	None	Low
Air System	None	Low
Wastewater Treatment	None	High
Flare and Incineration	None	Low
Tankage, Shipping, Receiving and Support Facilities	None	Low

[a]Based on gas input.

Air Separation

For purposes of the demonstration plant, the required scale-up factor is 1.3 to 1. This should not pose any problems since the principles of air separation are well understood and the techniques utilized in the design of such plants are well known.

Gasification

Although fixed-bed gasification of coal has been known for nearly a century, application of this technology to the gasification of strongly caking coals is a new field presently being explored. Experience on the gasification of this type of coal is reduced to a few runs and limited to small pilot units.

Insights gained from the gasification program at Westfield indicate that the process can be successfully carried out. Areas which may pose problems in the demonstration plant are (1) the size of the new gasifier (scale-up factor of 2.8 to 1); (2) prevention of uncontrolled coke formation in the shaft of the gasifier above the tuyeres at high throughputs; (3) formation of free iron in the slag pool on the hearth of the gasifier and splattering of slag in the slag quench vessel so that it adheres to the quench vessel internals and plugs it, impeding the flow of slag; and (4) operating life of the materials of construction at the bottom of the gasifier.

Based on operating experience with the gasifier unit at Westfield, the transition to the demonstration size unit is expected to be free of insurmountable problems.

Early in the operation of the 300-ton/day gasifier, the problem of coke formation in the shaft of the gasifier was solved by extending the stirrer in the upper section of the gasifier shaft to provide better gas solid contacting, and by operating the stirrer at speeds more suitable to the bed conditions. The effect of these two changes is to improve the performance of the gasifier by enhancing the probability that the coal particles would pass through their caking phase and be broken up into smaller, more manageable bits before reaching the raceway region.

Blast furnaces utilized in the steel industry operate in much the same way as the gasifier and problems in the hearth are similar. The amount of iron in a blast furnace is much greater than in the gasifier. The blast furnace relies on water cooled tuyeres and slag notch, and cooled high conductivitiy refractories and obtains reasonable commercial life from its components. The internal parts of the gasifier operating at Westfield performed satisfactorily during the runs, with only one component, the slag tap, failing during one of the early runs. The failure was due to poor bed conditions in front of the tuyeres. Table 6-18 shows the number of hours of operation and the condition of the various components at the end of the trials. The different number of operating hours for the different components reflect changes made in the gasifier to cope with problems encountered with the use of Eastern U.S. caking coals.

Shift Conversion

This process has been extensively used in other commercial applications. A prototype unit was tested at Westfield in conjunction with the gasifier program; no major problems were found in the operation of the unit.

The reactor operation is based on the reaction of carbon monoxide and steam over a catalyst bed.

$$CO + H_2O \rightarrow CO_2 + H_2$$

172 GASIFICATION

A problem which could arise at the magnified conditions of the commercial plant is the severe fouling of the catalyst bed due to a poor supply of hydrogen in the feed. Hydrogen is used to saturate the heavier unsaturated hydrocarbons. If not enough hydrogen is present, the heavier hydrocarbons could polymerize and completely plug the reactor.

Gas Cooling

Heat exchangers are designed and manufactured for a wide variety of applications, some of which are more demanding than what would be required for the demonstration plant.

Rectisol

This process has been used in the hydrocarbon processing and ammonia production industry where the carbon dioxide (CO_2) and hydrogen sulfide (H_2S) levels in the feed streams must be extremely low to satisfy the process purity requirements. Units have been built which can process up to 150,000 scf/day of feed gas. No problems are expected from the operation of the Rectisol units at the scale-up size required for the demonstration plant.

Methanation

This section utilizes a methanation unit of the type used in naphtha-to-SNG plants which has been modified to operate on synthesis gas. Feed to the unit consists of hydrogen, carbon monoxide and nonmethane hydrocarbons. Hydrogen reacts with the carbon monoxide and unsaturated hydrocarbons over a nickel catalyst to form a methane rich gas.

A semicommercial plant operated at Westfield in conjunction with the gasification program was operated on a feed of synthesis gas. These trials proved conclusively the ability of the process to produce high-Btu pipeline gas.

Drying and Compressing

This section consists entirely of well proven technology and has been in use for many years in the gas industry.

Sulfur Recovery

This section relies on Claus sulfur recovery plants which are very common throughout the hydrocarbon processing industry.

MAJOR GASIFICATION PROCESSES 173

Slag Handling

The technology is similar to that used in coal-fired power plants.

Gas Liquor Separation

The required scale-up factor for purposes of the demonstration plant is 1.3 to 1. The unit operates on gravity separation of the stream components. The technology is well known and has been extensively used in the chemical industry. Some of the users include the coal gasification plant at the Sasol complex in South Africa.

Phenol Extraction

The Lurgi Phenosolvan process has extensive commercial applications, one of which is the coal gasification plant at the Sasol complex in South Africa. For purposes of the demonstration plant, the unit is required to process about 1×10^6 lb/hr (0.454×10^6 kg/hr) of gas liquor, which is within the capabilities of the phenol extraction unit (1.1×10^6 lb/hr or more).

Ammonia Recovery

The Chemie-Linz-Lurgi ammonia recovery process is part of the Lurgi Phenosolvan process and is extensively used in the industry. It is a proven process and provides a feasible method of producing anhydrous ammonia from dilute feeds.

Water Treatment and Steam Generation

The water treatment and steam generation uses well proven conventional equipment at proven throughput capacities. The Wellman-Lord flue gas desulfurization has been found to be reliable in many operations in the hydrocarbon processing industry.

Cooling Water System

This is a well known process. Cooling requirements from the plant are well within present equipment capabilities.

Plant and Instrument Air Systems

This is a well known, commercially reliable technology.

Wastewater Treatment

Processing of industrial wastewater is a well known technology and is relatively easy to apply when the waste stream components are known. In a coal gasification plant environment this is not the case, mainly because the different potentially polluting water streams have not been characterized accurately as to type and concentration of each component. The problems are compounded by the fact that complete characterization of the streams can only be made from data from the operating plant. In the design of the demonstration plant a number of assumptions were made concerning the wastewaters. These assumptions were made based on information available from processes similar to those within the plant and from the performance of the individual units which would be part of the plant. These estimates, however, could well be inaccurate.

Another problem for which there exists a good deal of uncertainty is the problem of disposing of the salt removed from the raw water that enters the plant. A chemical fixation is proposed to microencapsulate the salts prior to burial in a landfill area. The process has been successfully demonstrated in other applications (i.e., fixation of spent caustic); however, its application in this particular case has not been demonstrated. The success of chemical fixation is dependent upon the characteristics of the salts and, therefore, leachability is uncertain without pilot studies.

Flare and Incineration

These facilities are of standard design and there is long experience in their operation.

Tankage, Shipping, Receiving and Support Facilities

These all are standard commercial areas.

Modification of Plant Design to Produce Medium-Btu Gas

Because the commercial plant design presented in this section represents the results of a design that has optimized process flows to produce a product for a specific application, alteration of the design to produce medium-Btu gas requires numerous subtle plant modifications. However, a general set of comments can be made regarding alteration. The shift conversion unit may be eliminated if a low ratio H_2/CO gas is acceptable. The Rectisol purification unit removes CO_2 and produces a very low sulfur level in the clean gas. Depending on the application a less expensive cleanup process could be substituted that would not remove the CO_2 and would leave more sulfur in the synthesis gas. A different cleanup system would also modify the Claus sulfur recovery plant and auxiliary processes. The methanator would be eliminated. The compressor and dryer sections would be affected in accordance with the uses of the gas. If a lower pressure were desired, energy could be recovered from a gas expander. Plant steam and cooling water flows will be changed in a medium Btu configuration.

Economic Analysis

Projected costs for the slagging Lurgi process are derived from studies of two plants. The largest is a commercial scale plant producing 250×10^6 scf/day of synthetic pipeline gas. The smaller is a demonstration plant producing 60×10^6 scf/day that uses gasifiers and other components of commercial plant size but has only 4 gasifiers instead of the 12 in the commercial plant. The comparative economics of gas produced by the two plants is displayed in Table 6-22. The cost of building the demonstration plant is $235 million (first quarter 1978).

The following analysis is based on the studies of the commercial plant design. Although coal gasification is a rather old process, commercial-scale conversion of coal to SNG has never been attempted before. The economic evaluation of the high-Btu gasification plant proposed by Conoco has been done based on operating performance data and actual costs for most of the components.

Table 6-22. Summary of Program Costs[a] ($ 1975)

Cost Element	Commercial Plant	Demonstration Plant
	(Costs $/$10^6$ Btu)	
Coal	1.185	1.029
Operation and Maintenance	0.714	1.310
Administration	0.066	0.163
Subtotal	1.955	2.502
Depreciation and Return on Investment	1.085	1.529
Income Taxes	0.661	0.998
Total Costs	3.701	5.029
By-product Credits	(0.243)	(0.243)
Net Cost of Gas	3.458	4.786

[a]Source: Sudbury, 1978.

GASIFICATION

Table 6-23. Overall Heat and Material Balance

	Mass Flowrate		Gross Heating Value		% of Heat Value
	lb/hr	kg/hr	10^6 Btu/hr	10^3 MJ/hr	
Input					
Coal to Gasification	1,406,566	638,018	14,332.50	15,114	79.00
Coal to Boilers	374,000	169,646	3,810.95	4,019	21.00
Excess Coal Fine	563,707	255,697			
Total Coal Input	2,344,273	1,063,362	18,143.45	19,134	100.00
Flux	69,154	31,368			
Air to Air Separation Plant	3,182,671	1,443,659			
Raw Water	6,102,022	2,767,877			
Combustion Air	4,289,242	1,945,600			
Chemicals	7,721	3,502			
Total Input	15,995,083	7,255,369	18,143.45	19,134	100.00
Output					
Products					
Pipeline Gas	430,325	195,195	9,666.79	10,195	53.28
Naphtha	14,988	6,798	265.64	280	1.46

Oil	21,873	9,921	373.24	394	2.06
Crude Phenols	5,712	2,591	73.29	77	0.40
Anhydrous Ammonia	4,010	1,818	38.77	41	0.21
Sulfur	76,566	34,730	307.40	324	1.70
Sodium Sulfate Purge	3,935	1,784			
Coal Fines	563,707	255,697			
Products Subtotal	1,121,016	508,493	10,725.13	11,310	59.11
Waste and Vent Streams					
Air Separation Plant Vents	2,443,583	1,108,409			
Combustion and Dryer Vents	6,487,821	2,942,875			
Cooling Tower and Steam System Water to Atmosphere	5,475,300	2,483,596			
Slag to Landfill	251,451	114,058	2.42		0.01
Misc. Waste Solids to Landfill	123,140	155,856			
Net Water Loss	92,772	42,081			
Vents and Wastes Subtotal	14,874,067	6,745,515	2.45	2.55	2.55
Heat Loss to Air Cooling			2,120.55	2,236	11.69
Evaporative Cooling and Heat Loss			5,246.36	6,633	28.92
Heat of Flux Calculation			48.99	52	0.27
Total Output	15,995,083	7,255,369	18,143.45	19,134	100.00

178 GASIFICATION

One key component within the plant, the gasifier, has only been operated at the pilot-plant scale. This is a unique unit which has been specifically designed for this application and for which there is no commercial counterpart. Data from this unit have been extrapolated from the pilot plant data to fit the requirements of the demonstration plant. The plant performance and feedstock data have been obtained from the conceptual design of the demonstration plant. The plant's overall heat and material balances are shown in Table 2-23.

The plant requires 2640 acres of land, with 480 acres for actual plant operations, and the remaining 2160 acres reserved for the buffer zone and waste solids disposal area.

Construction of the plant is assumed to be completed over a four-year period, with plant life estimated at 20 years. The erected plant cost is estimated to be $918,815 million. Excluded from this estimate are the cost of the land; catalysts and chemicals; startup expenses; craft training program; royalties and license fees; financing charges; working capital; contractor's home office, overhead and profit; and contingency for appropriations.

The estimate is based on material and labor price structure as of the first quarter 1978 without escalation. A detailed breakdown of the estimate on a section by section basis is shown in Table 6-24. The costs shown for each section include construction of all facilities within the battery limits of that section. Facilities outside the battery limits of each section, such as interconnecting pipeway, buildings, plant roads and utility distribution systems are reported separately under support facilities.

The estimates have been broken into three different categories: material, subcontract and labor. The material section includes process equipment and other materials of construction. The subcontract category includes major process and utility subsystems, such as air separation plant, coal handling equipment and steam generation, and civil work, insulation, painting and fireproofing. The labor costs refer to direct–field labor costs, and are based on first-quarter labor contracts in effect in the Wood River, IL, area, which is the expected site of the demonstration plant.

The plant thermal efficiency is 59.1%. The thermal efficiency is defined as the ratio of the total energy output, i.e., the energy contained in the product gas, the by-products and the steam generated by the process, to the total energy contained in the coal feedstock, the steam and the electricity utilized in the process.

Two methods of financing are used in this analysis, the discounted

MAJOR GASIFICATION PROCESSES 179

cashflow method (DCF) and the public utility method. The results under the DCF will be discussed first.

In the discounted cash flow (DCF) method the following assumptions are made.

- plant operating life: 20 years
- depreciable life: 16 years
- depreciation type: sum-of-the-years-digits
- DCF rate of return: 12%
- income tax rate: 48%
- coal cost: $1.00/$10^6$ Btu

Under this method the total capital investment is $1.3 billion, including $146 million for contingency, $76 mililon for working capital and $50 million for startup costs. Gas costs are estimated at $6.60/$10^6$ Btu of synthetic pipeline gas. When the assumptions are changed to zero federal income tax burden and a DCF of 9%, the price is $4.85/$10^6$ Btu of gas produced.

The total capital investment when utilizing Illinois No. 6 coal is shown in Table 6-25. If the land acquisition cost is included in the first year's total, the startup cost is allocated to the fourth year, and the working capital distributed over the four-year construction period, then 12.3% of the total investment is spent during the first year. The second and third year account for 33.9 and 39%, respectively. The expense during the fourth year is 14.8%.

Table 6-26 shows the total capital investments when using alternative feedstock, Ohio No. 9 and Pittsburgh No. 8 coals. For ease of comparison, the associated costs when utilizing Illinois No. 6 coal are also presented in the same format.

The plant's annual net operating cost of $215 million is shown in Table 6-27. This is calculated from a gross operating cost of $286 million, less a credit of $71 million for the sale of by-products. A summary of the raw materials inventory and normal usage is presented in Table 6-28. The raw material inventory is based on storage capacity. The inventory for raw coal is maintained at 840,000 tons, as coal is continuously received from the mine. This will supply the plant for 30 days at the normal operating rate. A maximum of a seven-day supply of flux and enough coke to start up 12 gasifiers are maintained in storage. A rate of 28,131 and 830 ton/day has been estimated for coal and flux usage, respectively.

180 GASIFICATION

Table 6-24. Plant Cost Summary ($1000s)

Section No.	Description	Direct Materials	Direct Labor w/o Productivity	Subcontracted Packages	Total Cost
100C	Coal and flux handling and preparation			32,670	32,670
200C	Air separation			106,700	106,700
300C	Gasification	39,925	4,181	2,848	46,954
400C	Shift conversion	39,013	9,140	3,623	51,776
500C	Gas cooling	6,421	1,704	921	9,046
600C	Rectisol	86,104	20,039	8,993	115,136
700C	Methanation	21,565	4,369	2,858	28,792
800C	Product gas compression and drying	10,117	1,808	1,170	13,095
900C	Sulfur recovery	4,618	1,317	806	6,741
1000C	Slag handling and disposal			2,650	2,650
1100C	Gas liquor separation	5,568	1,940	4,915	12,423
1200C	Phenol extraction	4,311	1,056	825	6,192
1300C	Ammonia recovery	5,494	1,481	751	7,726
2000C	Water treatment and steam generation	8,174	1,463	173,263	182,900
2400C	Cooling water system	4,575	2,880	7,408	14,863

2500C	Plant and instrument air system	357	87	23	467
2700C	Wastewater treatment	1,690	1,229	21,656	24,575
3000C	Flare and incinerator facilities	7,134	2,100	1,746	10,980
3100C	Tankage and fuel oil system	1,272	515	2,341	4,128
3200C	Shipping and receiving facilities	646	325	163	1,134
4000C	Support facilities	26,038	10,127	17,743	53,908
	Subtotal	273,022	65,761	394,073	732,856
	Construction indirects				85,490
	Labor productivity				75,625
	Spare parts				3,200
	Soil consultant				80
	Environmental consultant				1,350
	All risk insurance				9,100
	Contractor's bond				5,150
	Sales/use tax				6,000
	Total Erected Plant Cost				918,851

Table 6-25. Capital Investment[a]

	Capital Cost (10^6 $)				
	1st Year	2nd Year	3rd Year	4th Year	Total
Erected Plant Cost	80.248	341.657	410.930	86.016	918.851
A&E Contractor's Profit	2.601	1.344	0.303	0.087	4.335
A&E Engineering and Design Cost	23.445	12.113	2.735	0.781	39.074
Lurgi Engineering and Design Cost	4.889	2.526	0.571	0.163	8.149
Subtotal	111.183	357.640	414.539	87.047	970.409
Project Contingency (15% of subtotal)	11.645	53.858	65.502	14.556	145.561
Offeror's Administration and Construction Engineering	1.561	1.514	1.324	0.331	4.730
Subtotal	124.389	413.012	481.365	101.934	1,120.700
Paid-up Royalties	10.389	6.916	5.533	5.639	28.641
Initial Charge of Catalysts and Chemicals				14.338	14.338
Total Plant Investment	134.942	419.928	486.898	121.911	1,163.679
Startup Cost					50.508
Land Aquisition Cost					5.280
Working Capital					76.142
Total Capital Requirement					1,295.609

[a] The engineering costs are estimated based on expected rate of completion of engineering during construction. All engineering charges which accrued prior to the start of construction are included in year one. The accuracy of the estimate is 25 percent.

MAJOR GASIFICATION PROCESSES 183

Table 6-26. SNG Product Gas Cost Alternative Coal Comparison

	Illinois No. 6		Ohio No. 9		Pittsburgh No. 8	
	10^6 \$/yr	\$/10^6 Btu	10^6 \$/yr	\$/10^6 Btu	10^6 \$/yr	\$/10^6 Btu
Coal Costs	189.192	2.471	191.495	2.500	184.318	2.407
Other Raw Materials	6.008	0.078	8.601	0.114	3.826	0.050
Catalysts and Chemicals	15.275	0.200	15.275	0.200	15.275	0.200
Labor and Benefits	16.996	0.222	16.996	0.222	16.996	0.222
Administration and General Overhead	3.399	0.044	3.399	0.044	3.399	0.044
Maintenance and Operating Supplies	30.993	0.405	30.255	0.394	30.503	0.399
Local Taxes and Insurance	23.274	0.304	22.720	0.297	22.906	0.299
Annual Royalties	0.999	0.013	0.999	0.013	0.999	0.013
Subtotal Gross Operating Costs	214.972	2.808	210.972	2.775	216.601	2.829
Interest on Land and Working Capital	7.329	0.095	7.269	0.095	7.196	0.094
Average Income Tax	108.124	1.412	105.601	1.379	106.449	1.390
Average Capitalization of Investment	175.318	2.290	171.201	2.236	172.585	2.254
Total Annual Cost	505.743		495.043		502.831	
Total SNG Price		6.605		6.465		6.567

184 GASIFICATION

Replaceable catalysts are used in three sections of the plant: the shift conversion unit, the methanation unit and the sulfur recovery unit. Under normal use, it is estimated that the shift section will require catalyst replacement once every three years, the methanation unit each year, and the sulfur recovery unit every two years. The catalyst and chemicals inventory requirements for the different units are presented in Tables 6-29 and 6-30.

The labor and benefits summary is shown in Table 6-31. By-product inventory and revenue is given in Table 6-32.

Table 6-27. Operating Cost (10^6 $, First Quarter 1978)

Raw Materials (Table 6-28)	192.872
Catalyst and Chemicals (Tables 6-29 and 6-30)	15.275
Utilities (water) 12,245 gal/min at $0.40/1000 gal	2.328
Labor and Benefits (Table 6-31)	16.996
Administration and General Overhead	3.399
Supplies	
Operating	5.818
Maintenance and Contracts	25.175
Local Taxes and Insurance	23.274
Annual Royalties	0.999
Total Gross Operating Cost Per Year	286.136
By-product Credit (Table 6-32)	(71.164)
Total New Operating Cost Per Year	214.972

Table 6-28. Raw Material Summary

	Cost ($/ton)	Inventory		Usage		
		tons	10^6 $	ton/day	10^3 $/day	10^6 $/yr
Raw Coal	20.38	840.000	17.119	28,131	573.310	189.192
Flux	13.00	2,500	0.033	830	10.790	3.561
Coke	60.00	600	0.036	6	0.360	0.119
Total			17.188		584.460	192.872

Table 6-29. Catalyst Summary

	Initial Fill (10^6 $)	Inventory (10^6 $)	Usage (10^6 $/yr)
Shift Catalyst	8.19	2.05	2.73
Methanation Catalyst	3.81	1.27	3.81
Sulfur Catalyst	0.4	0.13	0.2
Total	12.4	3.45	6.74

MAJOR GASIFICATION PROCESSES

The cost of the product gas has been calculated for two different cases; they differ from each other in the income tax rate and the DCF rate of return. Case A is the 9% DCF rate of return and 0% income tax case, Base Case B is the 12% DCF and 48% income tax case. For each case, a sensitivity analysis showing the variation in gas price with coal costs, DCF rate of return, operating cost, and capital investment has been prepared. These are shown in Figures 6-8 to 6-12, respectively.

Table 6-33 shows the breakdown of the cost of the gas produced. Illinois No. 6 coal is considered under both cases A and B. Gas cost associated with the utilization of Ohio No. 9 and Pittsburgh No. 8 coals is presented under case B.

The public utility financing is different from the DCF method in several aspects. The public utility financing assumes the following:

1. The major portion of the required capital is borrowed, with accrued interest added to the total capital requirements.
2. After construction, the public utility invests equity capital. The public utility then receives from the plant operation a return on its equity and the lending institution is paid interest on the loan.
3. Depreciation charged to the project reduces the debt and equity simultaneously. Allocation of depreciation to retirement of debt and equity is determined by the initial debt-equity ratio.

Total capital requirements under this method of financing are shown in Table 6-4.

The total maintenance for the plant is based on plant investment. A fraction of the investment for each section is taken as an annual maintenance expense. The fraction used varies from section to section. Six percent is used for sections requiring extensive solid handling, three percent is assessed for other onsite units, with one percent for other offsite units. A detailed maintenance breakdown is presented in Table 6-35. The totals shown in the table cover both labor and supplies, with 60% of the total allocated to labor, and 40% to supplies.

The plant annual net operating cost of $217 million is shown in Table 6-36. This is calculated from a gross operating cost of $288 million, less a credit of $71 million for the sale of by-products. The cost of the product gas for the first year of operation and the 20-year average price is shown in Table 6-37.

A sensitivity analysis showing the variations in gas price with capital investment, coal cost, debt interest rate, maintenance and operating expense, by-product value and plant yield variations is shown in Figures 6-13 through 6-17, respectively.

Table 6-30. Chemical Summary

	Units	Cost ($/unit)	Initial Fill Capacity (units)	Initial Fill Cost ($1000s)	Inventory Capacity (units)	Inventory Cost ($1000s)	Usage Units/Hour	Usage $1000/Year
Methanol	lb	0.053	4,300,000	227,900	280,000	14,840	700	293.830
Caustic (5 wt %)	lb	0.078			260,000	20.150	3,000	1,841.400
Triethylene Glycol	gal	4.404	2,700	11,890	1,500	6.610	1.04	36.270
Isopropyl Ether	lb	0.22	1,200,000	264,000	20,000	4.400	6	10.450
Propylene	gal	0.50	122,000	61,000	10,000	5.000	1	4.000
Fuel Oil	gal	0.36	3,360,000	1,209,600	100,000	36.000	13[a]	37.070
Silica Gel Desiccant	lb	0.64	12,000	7,680				
Sulfuric Acid	lb	0.028	300,000	8,550	300,000	8.550	1,050	237.000
Dust Suppression Chemical	gal	5.80			2,000	11.600	0.53	24.360
Raw Water Chemicals								
Alum (5 wt %)	lb	0.04			180,000	7.200	125	39.600
Polymer	lb	2.28			15,000	34.200	10	180.580
Lime	lb	0.015			860,000	12.900	600	71.280

MAJOR GASIFICATION PROCESSES 187

Chlorine	lb	0.12	6,000	0.720	4	3.800
Trisodium Phosphate	lb	0.48	32,000	15.360	22	83.640
Morpholine (40%)	lb	1.03	29,000	29.870	20	163.150
Hydrazine (35%)	lb	2.00	700	1.400	0.5	7.920
Soda Ash	lb	0.04	4,000,000	176.000	3,043	964.020
Cooling Tower Chemicals						
Chlorine	lb	0.12	18,000	2.160	12.5	11.880
Dispersant	lb	1.05	100,000	105.000	70.8	588.770
Chrome Inhibitor	lb	0.60	60,000	36.000	41.7	198.160
Zinc Inhibitor	lb	0.60	18,000	10.800	12.5	59.400
Wastewater Chemicals						
Lime	lb	0.015	400,000	6.000	275	32.670
Polymer	lb	2.28	15,000	34.200	10	180.580
Phosphoric Acid (54%)	lb	0.18	100,000	18.000	70	99.790
Fixation Chemical	lb	0.233	2,600,000	605.800	1,800	3,321.650
Activated Carbon	lb	0.49	16,000	7.840	11.2	43.460
Total			300,000	147,000 1,937.620 1,210.600		8,534.730

[a]Yearly average.

188 GASIFICATION

Table 6-31. Labor and Benefits Summary

	Number Required	Total Salary ($)
Plant Management	2	76,000
Administration	57	1,112,500
Technical Support	64	1,351,500
Operation and Maintenance Staff	57	1,459,500
Process Operations	276	5,288,400
Maintenance Mechanics	200	3,917,000
Subtotal	676	13,204,900
Process Operations Overtime (10% of $5,288,400)		528,800
Maintenance Mechanics Overtime (6% of $3,917,000)		235,000
Total Annual Payroll		13,968,700
Benefits (16% of $13,204,900)		2,112,800
Social Security and Unemployment Insurances		914,000
Total Salary Cost		16,995,500

Table 6-32. By-product Summary

Sulfur[a]	40.00	8,500	0.340	820.35	32.814	10.829
Ammonia	150.00	500	0.075	48.12	7.218	2.382
Tar-Oil	84.90	2,000	0.170	262.48	22.285	7.354
Phenols	160.00	600	0.096	68.54	10.966	3.619
Naphtha	101.76	1,500	0.153	179.86	18.303	6.040
Coal Fines	18.34	22,600	0.414	6,764.48	124.061	40.940
Total			1.248		215.647	71.164

[a] Long Tons.

Environmental Evaluation

Determination of the environmental effects of a large gasification plant is one of the principal concerns of proceeding with a synthetic fuel technology. While all processes have been designed to restrict emissions during routine operations, the methods to be used have not been demonstrated in large-scale commercial practice. Consequently, their effectiveness during normal operations must be determined by measurements on facilities yet to be constructed. The effect of process upsets and equipment malfunctions is undetermined. The routine process atmospheric emissions are presented in Table 6-38. The potential for pollution from different processes is shown in Table 6-39.

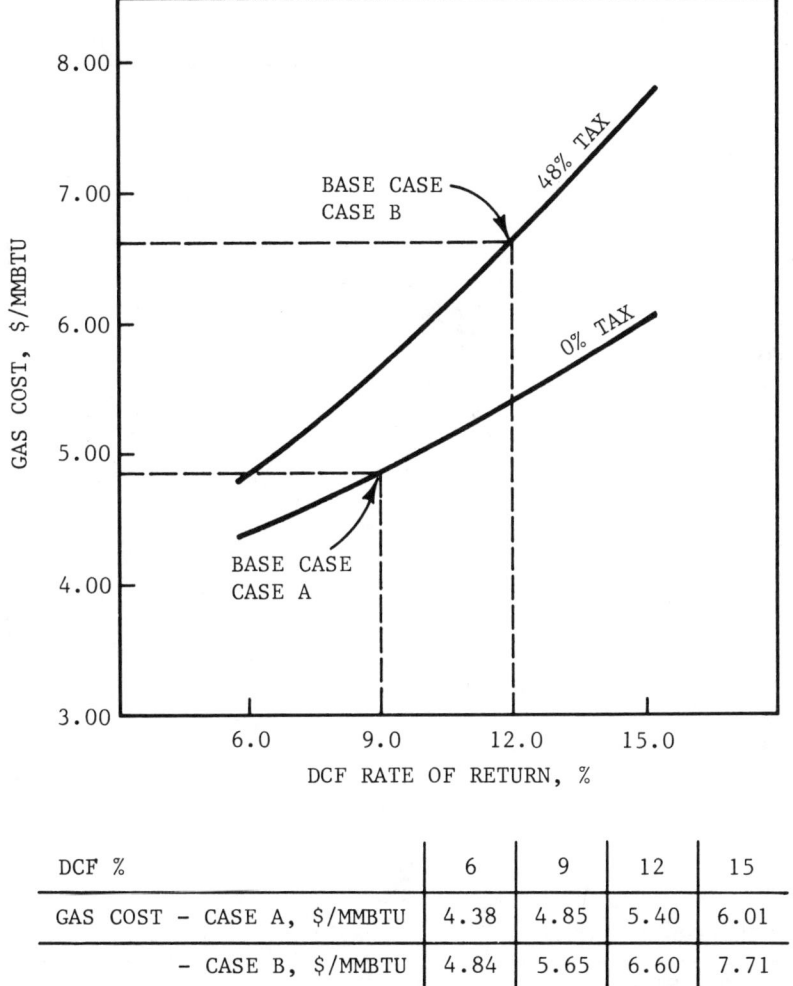

DCF %	6	9	12	15
GAS COST - CASE A, $/MMBTU	4.38	4.85	5.40	6.01
- CASE B, $/MMBTU	4.84	5.65	6.60	7.71

Figure 6-8. Gas cost variation with DCF rate of return.

Coal and Flux Handling Equipment

Because of the nature of coal mining operations, run-of-the-mine (ROM) coal has a high coal dust content, which is susceptible to being blown away by the wind or carried off by rainfall. The plant is to operate with a 30-day supply of ROM coal (about 840,000 tons of coal) and a 7-day standby supply of coal (about 200,000 tons of coal). The

190 GASIFICATION

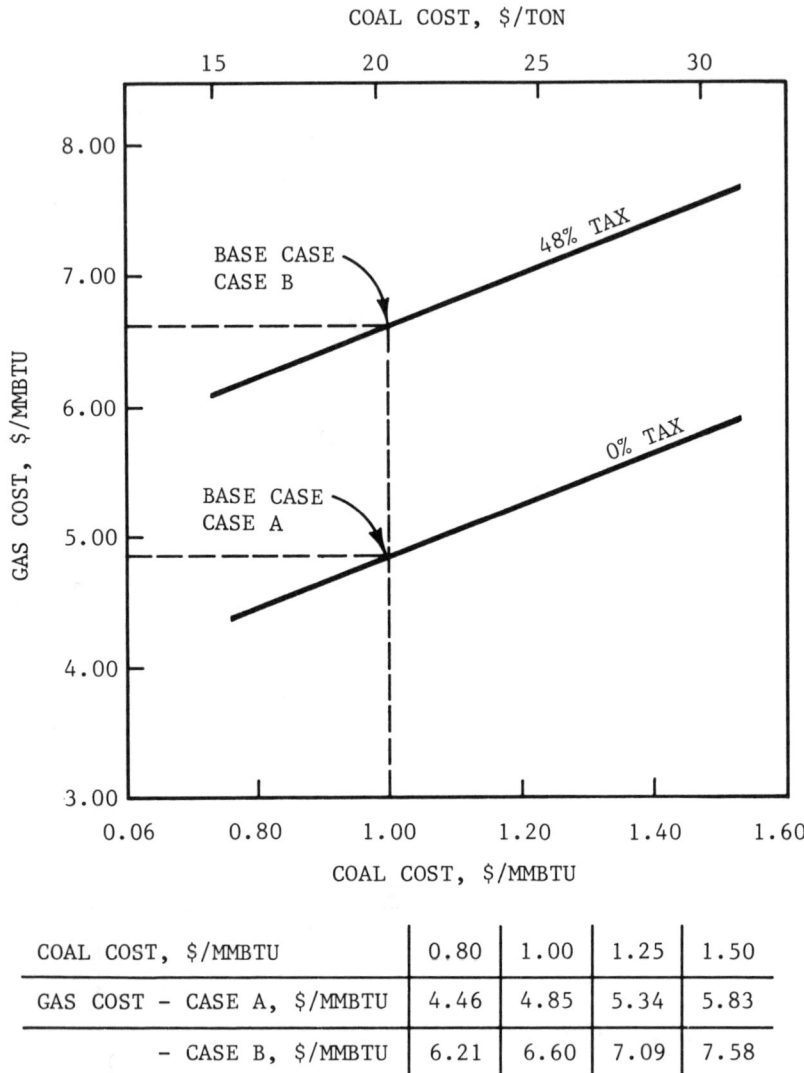

Figure 6-9. Gas cost variation with coal cost.

COAL COST, $/MMBTU	0.80	1.00	1.25	1.50
GAS COST - CASE A, $/MMBTU	4.46	4.85	5.34	5.83
- CASE B, $/MMBTU	6.21	6.60	7.09	7.58

30-day supply pile is to be used continuously. The 7-day pile is to be utilized only in case of normal supply cutoff; otherwise, it is to remain inactive.

All aspects of the coal transportation, from the mine to the storage area to the lock hoppers on the gasifier, are potential sources of coal

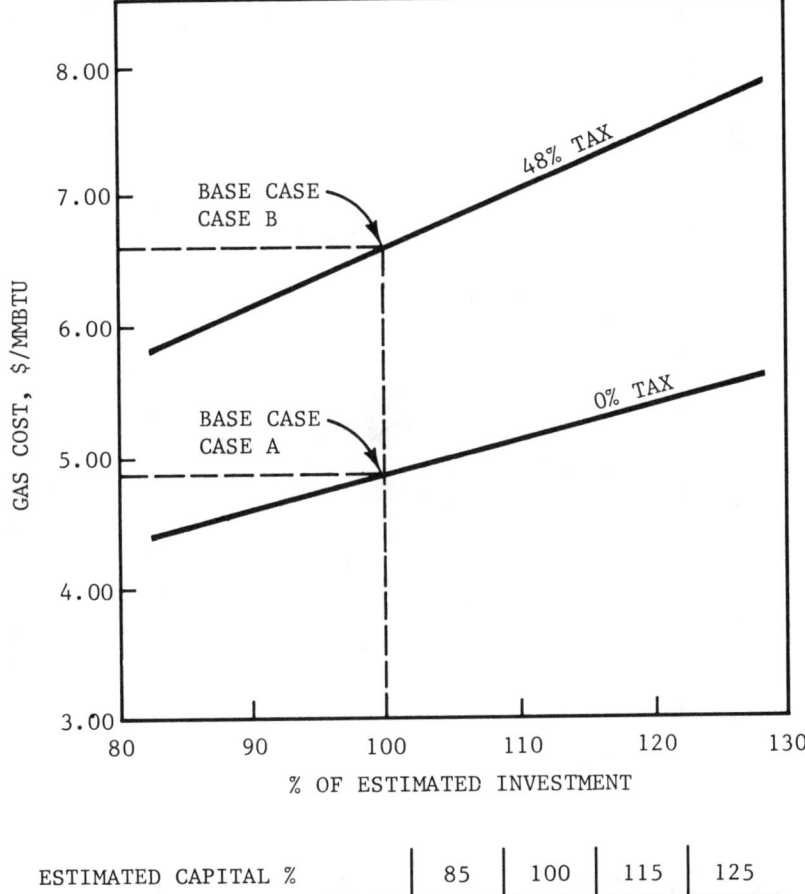

ESTIMATED CAPITAL %	85	100	115	125
GAS COST - CASE A, $/MMBTU	4.45	4.85	5.25	5.52
- CASE B, $/MMBTU	5.94	6.60	7.27	7.71

Figure 6-10. Gas cost variation with capital investment.

dust. The problem is alleviated to a great extent by mixing the mined coal with water before preparation and gasification.

Both coal storage areas are sources of contaminants. The 30-day pile of coal dust, along with the 7-day coal pile, are sources of water pollutants. The potential for air pollution comes from the wind removing dust particles from the surface of the pile. Water pollution potential

192 GASIFICATION

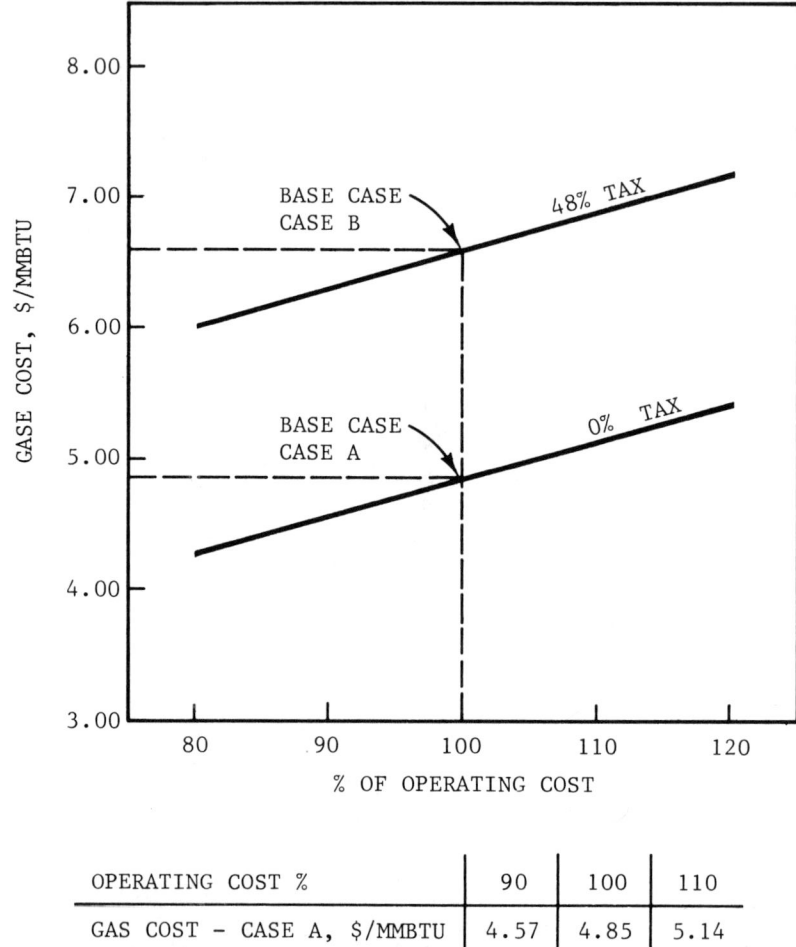

OPERATING COST %	90	100	110
GAS COST - CASE A, $/MMBTU	4.57	4.85	5.14
- CASE B, $/MMBTU	6.32	6.60	6.89

Figure 6-11. Gas cost variation with operating cost.

arises from the mechanical and chemical erosion by wind and rainfall. The 7-day pile will be compacted and coated with a polymer or asphalt coating to prevent dust and leachates from escaping the pile. The coat solves the coal pollution problems but brings some of its own. Chemical erosion on the coat surface will cause, depending on the material used, oils, greases and other asphalt material, or polymer material to

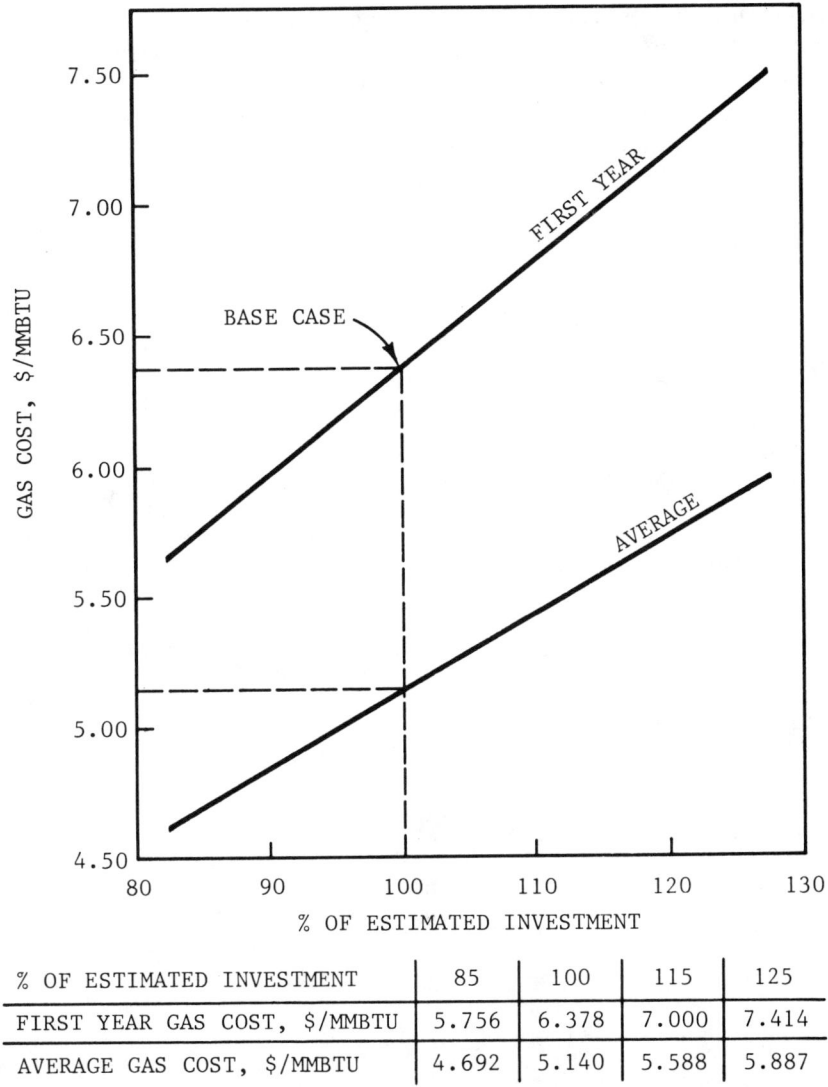

% OF ESTIMATED INVESTMENT	85	100	115	125
FIRST YEAR GAS COST, $/MMBTU	5.756	6.378	7.000	7.414
AVERAGE GAS COST, $/MMBTU	4.692	5.140	5.588	5.887

Figure 6-12. Gas cost variation with capital investment (utility financing economics).

be present in the water runoff. Soil erosion will be enhanced due to runoff. To prevent environmental damage, the rainwater runoff will be impounded and treated for reuse within the plant.

194 GASIFICATION

The storage of large quantities of mined coal in piles can create problems since it closely resembles conditions at the mine. Oxidation of the coal leading to spontaneous combustion can occur. The risks of spontaneous combustion in the coal pile are greatly reduced by the coating material since air is not allowed to go into the coal pile to oxidize the organic material.

Table 6-33. Product SNG Selling Price ($/$10^6$ Btu): DCF Method

	Base Case A Illinois No. 6[a]	Illinois No. 6[b]	Ohio No. 9 Coal[c]	Base Case B Pittsburgh No. 8 Coal[d]
Operating Costs				
Coal used in gasifiers	1.483	1.483	1.501	1.444
Coal used in boilers	0.394	0.394	0.366	0.394
Coal fines	0.594	0.594	0.633	0.569
Other raw materials	0.078	0.078	0.114	0.050
Catalysts and chemicals	0.200	0.200	0.200	0.200
Labor and benefits	0.222	0.222	0.222	0.222
Administration and general overhead	0.044	0.044	0.044	0.044
Maintenance and operating supplies	0.405	0.405	0.394	0.399
Local taxes and insurance	0.304	0.304	0.297	0.299
Annual royalties	0.013	0.013	0.013	0.013
Gross operating costs	3.737	3.737	3.784	3.634
Less Fines Credit	−0.534	−0.534	−0.570	−0.512
Less By-products Credit	−0.395	−0.395	−0.459	−0.293
Net Operating Cost	2.808	2.808	2.755	2.829
Interest on Land and Working Capital	0.074	0.095	0.095	0.094
Average Income Tax	0.000	1.412	1.379	1.390
Average Capitalization of Investment	1.969	2.290	2.236	2.254
Total SNG Cost	4.851	6.605	6.465	6.567

[a] GHV efficiency = 59.11%.
[b] GHV efficiency = 59.11%.
[c] GHV efficiency = 60.99%.
[d] GHV efficiency = 58.89%.

If the ground underneath the coal piles is permeable, additional protection will be installed beneath the coal piles to protect the groundwater. It should be noted, however, that in the event of failure of the natural or manmade barriers, the contamination of the groundwater will not be noticed after several years.

Table 6-34. Total Capital Requirement (Utility Financing Economics)

Total Plant Investment	918,851
Erected Plant Cost	51,588
Subtotal	970,409
Project Contingency (15% of subtotal)	145,561
Offeror's Overhead	4,730
Subtotal	1,120,700
Land (2,640 ac at $2,000/ac)	5,280
Total Plant Investment	1,125,980
Initial Charge of Catalysts and Chemicals	14,338
Paid-up Royalties	28,641
Allowance for Funds Used During Construction (total plant investment x 2 years x 9%)	202,676
Startup Cost (20% of total gross operating costs)	57,540
Working Capital	
Raw Materials (14-day supply)	8,182
Materials and Supplies (0.9% of total plant investment)	10,134
Net Receivables (0.5 month of gas, by-products)	23,313
Total Working Capital	41,629
Total Capital Requirements	1,470,804

Table 6-35. Total Maintenance Requirements, Utility Financing Economics ($1000s)

Plant Section		Direct Costs plus Sub-contracts	Percent Maintenance	Maintenance per Year
No.	Title			
100	Coal and flux handling and preparation	32,670	6	1,960
200	Air separation	106,700	3	3,201
300	Gasification	46,954	6	2,817
400	Shift conversion	51,776	3	1,553
500	Gas cooling	9,046	3	271
600	Rectisol and refrigeration	115,136	3	3,454
700	Methanation	28,792	3	864
800	Product gas compression and drying	13,095	3	393
900	Sulfur recovery unit	6,741	3	202
1000	Slag handling and disposal	2,650	6	159
1100	Gas liquor separation	12,423	3	373
1200	Phenol extraction			
1300	Ammonia recovery	7,726	3	232
2000	Water treatment and steam generation	182,900	3	5,487
2400	Cooling water system	14,863	1	149
2500	Plant and instrument air system	467	1	5
2700	Wastewater treatment	24,575	3	737
3000	Flare and incinerator facilities	10,980	1	110
3100	Tankage	4,128	1	41
3200	Shipping and receiving	1,134	1	11
4000	Support facilities	53,908	1	539
Total Maintenance				22,744

196 GASIFICATION

Table 6-36. Annual Operating Costs (Utility Financing Economics)

	Cost ($1000s)
Raw Materials	192,872
Catalysts and Chemical	15,275
Purchased Water (12,245 gal/min at 40¢/1,000 gal)	2,328
Labor	
Process Operating Labor (62 men/shift × 8,304 hr/yr × $9.21/hr)	4,742
Maintenance Labor (60% of total maintenance)	13,646
Supervision (20% of operating and maintenance labor)	3,678
Total Labor	22,066
Administration and General Overhead (60% of total labor)	13,240
Supplies	
Operating (30% of process operating labor)	1,423
Maintenance (40% of total maintenance)	9,098
Total Supplies	10,521
Annual Royalties	999
Local Taxes and Insurance (2.7% of total plant investment)	30,401
Total Gross Operating Cost per Year	287,702
By-product Credits	(10,829)
Sulfur	(2,382)
Ammonia	(7,354)
Tar Oil	(3,619)
Phenols	(6,040)
Naphtha	(40,940)
Total By-product Credit	(71,164)
Total Net Operating Cost per Year	216,538

Air Separation

No liquid or solids are emitted from this section. The only effluent is nitrogen gas which is vented to the atmosphere. Little or no environmental effects are associated with this section.

Gasification

This component has never been tested commercially as it has been expressly designed for this application. Its ability to perform under the conditions specified for the demonstration plant is yet to be proven. Emissions from the unit are prevented by the use of lock-hoppers to accept the feedstock (coal, limestone and recycled tars) and discharge the gasification wastes (slag grit and limestone). The intake lock-hopper is pressurized and continuously purged with carbon dioxide to

MAJOR GASIFICATION PROCESSES 197

prevent synthesis gas from entering the lock-hopper. After the coal discharge operation is completed, the pressurized gases in the lock hopper may contain up to 40 gr coal dust/100 scf of gas. To dispose of the dust, the gases are transported to the flare and incinerator facilities for incineration.

Two other gas streams are vented directly to the atmosphere. One is a low-pressure carbon dioxide gas stream from the input lock hoppers. This stream contains carbon dioxide and is passed through a cartridge filter to remove any entrained coal dust prior to venting. The other stream comes from the slag quench chamber; its composition is somewhat different, 90% carbon dioxide and 10% oxygen. No other emissions are expected from this unit.

Table 6-37. SNG Cost of Production (Utility Financing Economics)

	First Year of Service		Average Cost of Service	
	10^6 $/Yr	$/$10^6$ Btu	10^6 $/Yr	$/$10^6$ Btu
Coal Cost	189.192	2.471	189.192	2.471
Other Raw Materials	6.008	0.078	6.008	0.078
Catalyst and Chemicals	15.275	0.200	15.275	0.200
Labor	22.066	0.288	22.066	0.288
Administration and General Overhead	13.240	0.173	13.240	0.173
Maintenance and Operating Supplies	10.521	0.137	10.521	0.137
Local Taxes and Insurance	30.401	0.397	30.401	0.397
Annual Royalties	0.999	0.013	0.999	0.013
Gross Operating Cost	287.702	3.757	387.702	3.757
Less By-product Credits	−71.164	−0.929	−71.164	−0.929
Net Operating Cost	216.538	2.828	216.538	2.828
Income Taxes	49.676	0.649	26.177	0.342
Return on Equity and Depreciation	125.274	1.636	99.817	1.303
Interest on Debt	96.867	1.265	51.044	0.667
Annual Cost	488.355		393.576	
SNG Cost		6.378		5.140

Gas Production Rate = 241.7 MM SCFD 330 Days/Yr 960 Btu/SCF.
= 79,761 MM SCF Per Year 960 Btu/SCF.
= 76.571 × 10^{12} Btu/Year.

The liquid mixture of coal ashes and limestone is quenched in a water bath where the slag is immediately solidified. It is then passed to a lock hopper from which it is sent to the slag handling and disposal section. The slag quench water is a recirculating stream and acts as a seal to

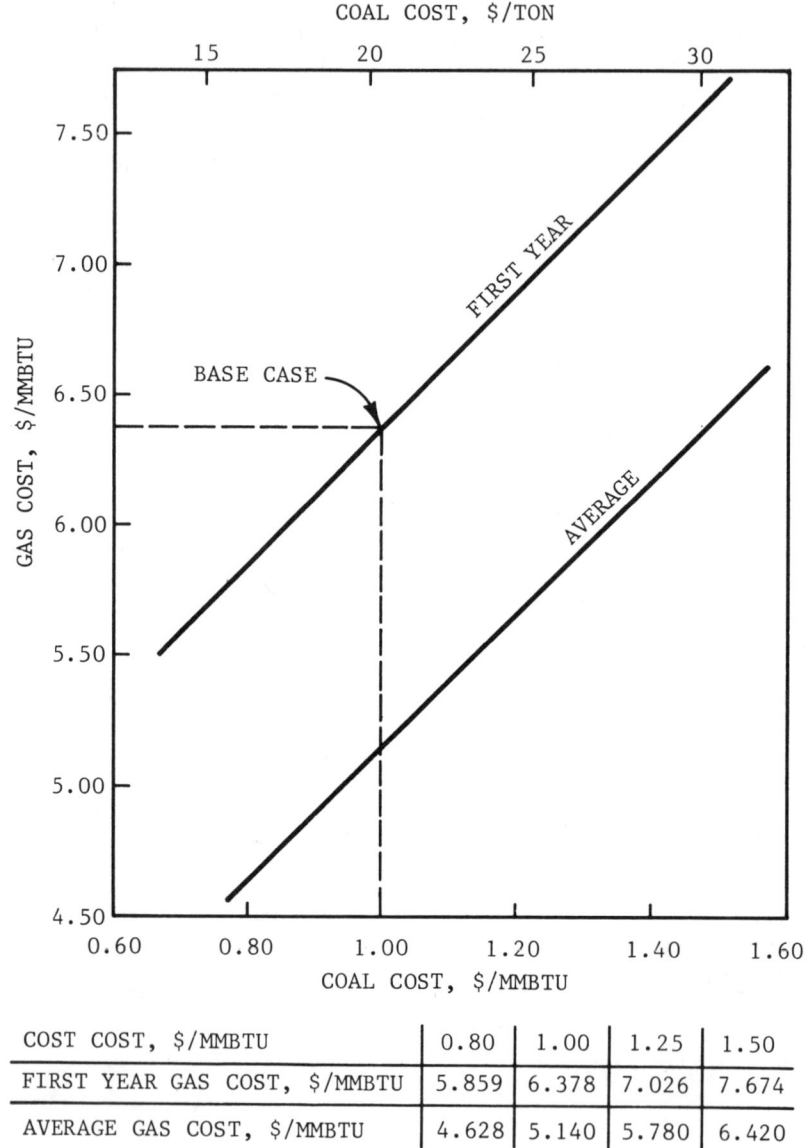

COST COST, $/MMBTU	0.80	1.00	1.25	1.50
FIRST YEAR GAS COST, $/MMBTU	5.859	6.378	7.026	7.674
AVERAGE GAS COST, $/MMBTU	4.628	5.140	5.780	6.420

Figure 6-13. Gas cost variation with coal cost (utility financing economics).

prevent the synthesis gas from entering the discharge lock hopper. Water that drained off with the slag is sent to the wastewater treatment section.

The hot product gases from the gasifier are quenched and scrubbed

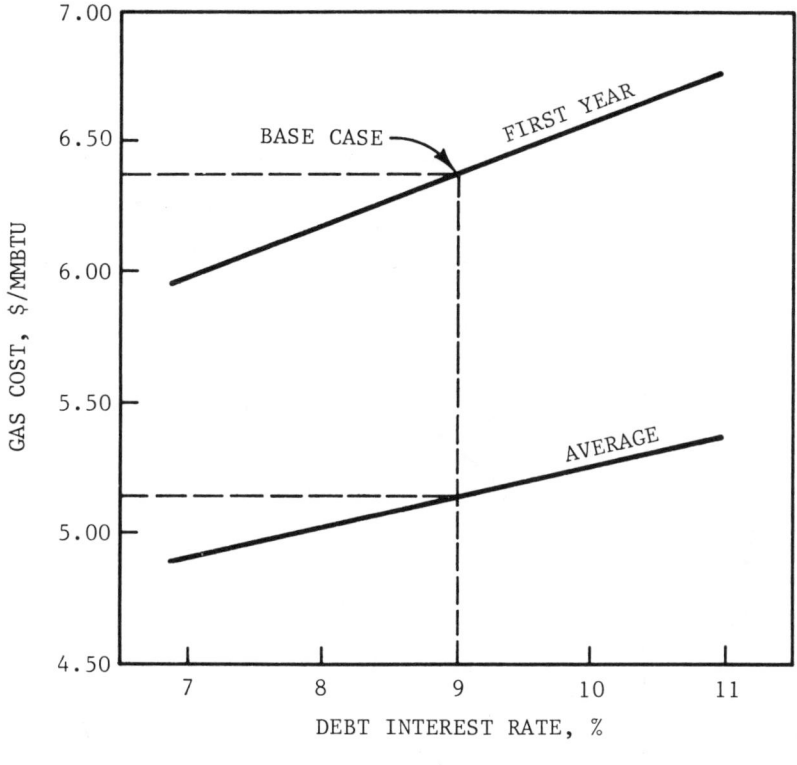

INTEREST RATE, %	8.0	9.0	10.0
FIRST YEAR GAS COST, $/MMBTU	6.184	6.378	6.576
AVERAGE GAS COST, $/MMBTU	5.031	5.140	5.252

Figure 6-14. Gas cost variation with debt interest rate (utility financing economics).

with circulating gas liquor stream, and passed to a heat exchanger where additional condensation of oils and other solids takes place.

Shift Conversion

This unit generates solid waste in the form of spent catalyst which is sent to the manufacturer for reactivation. No environmental effects are associated with this unit.

200 GASIFICATION

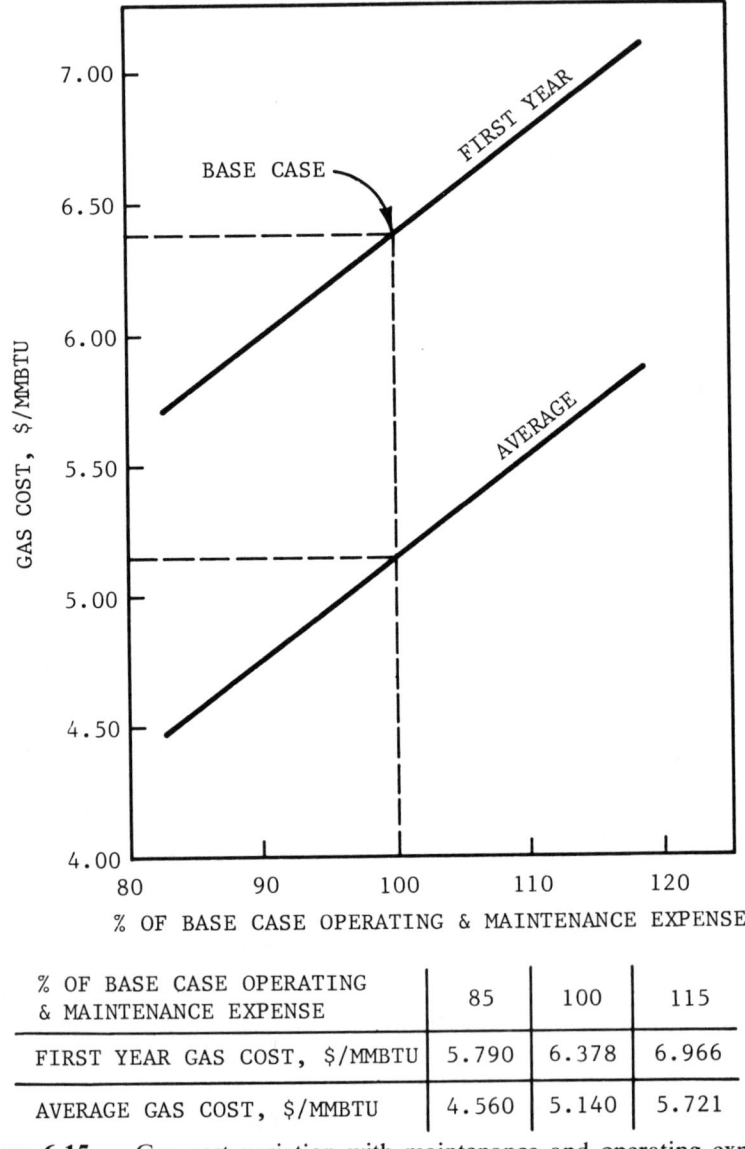

Figure 6-15. Gas cost variation with maintenance and operating expense (utility financing economics).

MAJOR GASIFICATION PROCESSES   201

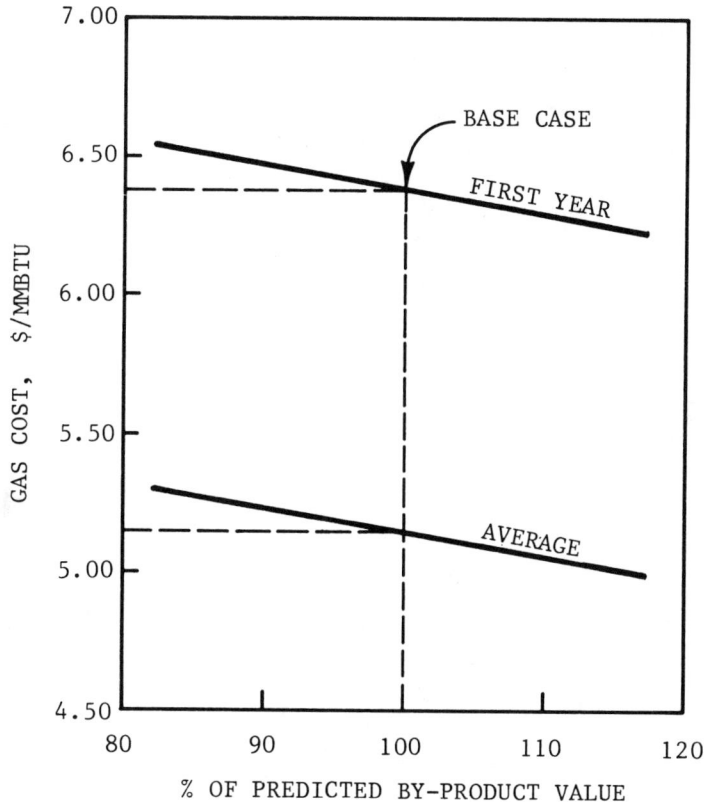

% OF PREDICTED BY-PRODUCT VALUE	81	100	115
FIRST YEAR GAS COST, $/MMBTU	6.517	6.378	6.238
AVERAGE GAS COST, $/MMBTU	5.279	5.140	5.001

Figure 6-16.   Gas cost variation with by-product value (utility financing economics).

Gas Cooling

Treated synthesis gas from the shift conversion unit is processed here prior to further processing in the Rectisol unit. No emissions are produced in this section.

202 GASIFICATION

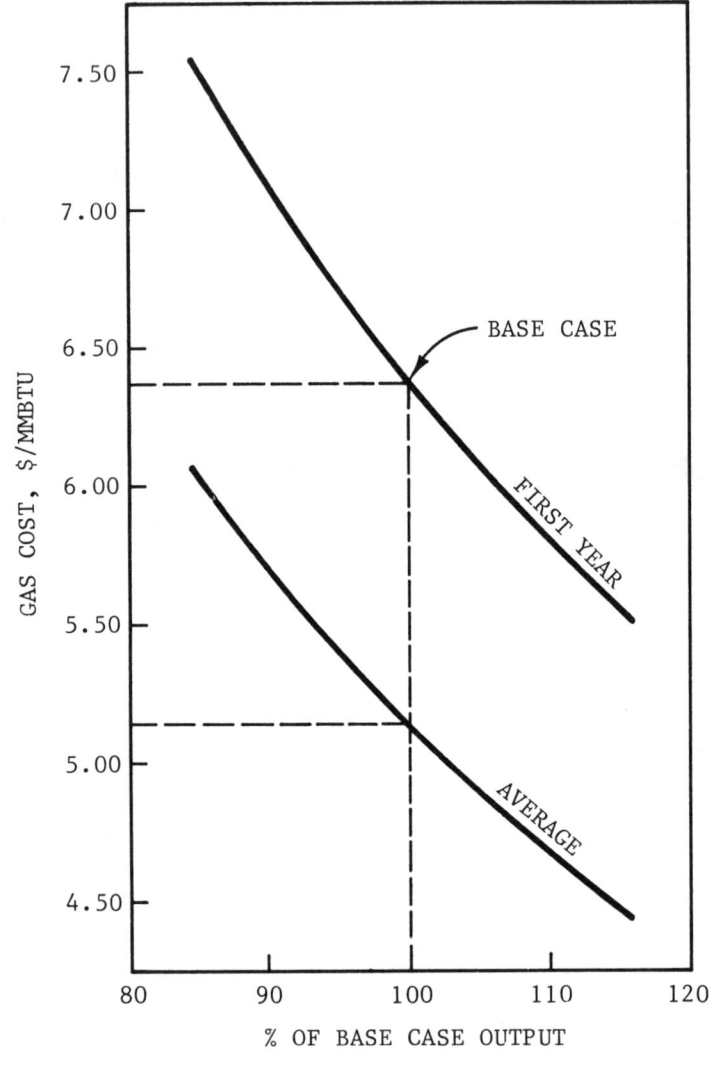

% OF BASE CASE OUTPUT	81	100	115
FIRST YEAR GAS COST, $/MMBTU	7.503	6.378	5.546
AVERAGE GAS COST, $/MMBTU	6.047	5.140	4.470

Figure 6-17. Gas cost variation with plant yield (utility financing economics).

Table 6-38. Process Atmospheric Emissions from the Commercial Plant (lb/hr)

	Air Separation	Lock Hoppers	Gasification Quench Vent	Gas Drying	Steam Generation			Flare	Total
					Stack Gas	Dryer Vent	Incinerator		
Nitrogen and Rare Gases	2,336,432		1		2,763,995	353,090	139,573		5,593,091
Oxygen	57,219		77		136,342	106,892	7,024		307,554
Carbon Dioxide		3,965	967		911,550	1,264	1,588,954	176	2,506,876
Sulfur Dioxide					2,356		346		2,702
Water Vapor[a]	19,782		116	534	407,878	38,393	17,540	7,044	491,297
Total	2,413,433	3,965	1,161	534	4,222,121	499,639	1,753,437	7,220	8,901,000
Temperature (°F)	94	68	250	220	180	120	350		

[a]The cooling water towers and steam system emit 5,475,300 lb/hr of water vapor to the atmosphere.

204   GASIFICATION

Table 6-39. Plant Component Pollution Potential During Normal Operations

Coal and Flux Handling and Preparation	Moderate
Air Separation	None
Gasification	Low
Shift Conversion	None
Gas Cooling	None
Rectisol	None
Methanation	Moderate
Product Gas Compression and Drying	None
Sulfur Recovery	Low
Slag Handling and Disposal	Moderate
Gas Liquor Separation	None
Phenol Extraction	None
Ammonia Recovery	None
Water Treatment and Steam Generation	Moderate
Cooling Water System	Low
Plant and Instrument Air Systems	None
Wastewater Treatment	Moderate
Flare and Incinerator Facilities	Moderate
Tankage and Fuel Oil Systems	Low
Shipping and Receiving Facilities	None

Rectisol

This section desulfurs the synthesis gas by contacting it with methanol. The solvent is regenerated by distillation and stripping and is recycled to the process. Several streams are produced in this section: a sulfur-rich stream, which goes to the sulfur recovery section; a naphtha by-product stream, which goes to the naphtha storage tank; a carbon dioxide-rich stream, which is used to pressurize the coal lock hoppers; a wastewater stream, which is sent to the wastewater treatment section; and a gas liquor stream, which is processed in the gas liquor separation section. There are no effluents released to the atmosphere from this section.

Methanation

This section has never been tested in the commercial production of SNG. Its biggest application has been at the pilot plant level, where it was operated for about two years prior to the DOE program.

The process offers the advantage of converting carbon dioxide (CO_2) and carbon monoxide (CO) to methane, with preference given to CO. Unsaturated hydrocarbons and alcohols are completely hydrogenated. It also achieves complete conversion of ammonia and nitric oxides to elemental nitrogen, while hydrogen cyanide is only partially destroyed.

MAJOR GASIFICATION PROCESSES 205

The formation of metal carbonyls is not expected to occur at the methanator operating temperatures, as the production of carbonyls is enhanced by lower temperatures; however, potentially favorable conditions are present during transient operations such as startup and shutdown of the individual units.

Product Gas Compression and Drying

The only effluent is water vapor, which is vented to the atmosphere. No environmental problems are associated with this unit.

Gas Liquor Separation

This section processes high-pressure condensed gas liquors and depressurizes them to atmospheric pressure to cause liquids and gases to separate. The gas is washed with gas liquor to recover ammonia. The gas is then passed to the sulfur recovery section. The depressurized liquor contains tar, tar oil, dust and impurities. These are gravity separated and disposed (the tar oil is sold). The other streams are returned to the gasification section for injection to the gasifier. After separation the gas liquor is sent to the phenol extraction section.

Phenol Extraction

Phenolic components are recovered from the gas liquor by extraction with isopropyl ether in a series of mixer-settler tanks. The gas liquor is sent to the ammonia recovery section from which it is sent back with sour gas produced in the recovery of ammonia. The sour gas is washed with phenol and sent to the sulfur recovery unit. No effluents are released to the atmosphere.

Ammonia Recovery

This section processes the dephenolized gas liquor to recover anhydrous ammonia. The stripped gas liquor is then returned to the phenol extraction unit. There are no effluents discharged to the environment from this unit.

Sulfur Recovery

The removal of sulfur compounds from the different gas streams converging to this section is accomplished in a Claus reactor by reacting the sulfur dioxide and hydrogen sulfide contained in the different streams. The main products of the reaction are water and elemental sulfur. The water is sent to the wastewater treatment section. The sulfur is collected and sold.

To obtain the proper mole ratio required for a complete destruction of the reactants, some hydrogen sulfide is oxidized to sulfur dioxide:

$$2H_2S + 3O_2 \rightarrow 2SO_2 + 2H_2O$$

which then reacts with hydrogen sulfide in the annihilation reaction

$$2SO_2 + 4H_2S \rightarrow 3S_2 + 4H_2O$$

to ensure complete removal of sulfur-bearing compounds. The tail gas from this section is sent to the steam generation and distribution component of water treatment and steam generation.

Water Treatment and Steam Generation

This section consists of three subsystems: water treatment, steam generation, and distribution and flue gas desulfurization. The water treatment plant will process raw water for use in the plant. The raw water is first sent to a clarifier where suspended solids are removed; it is then treated with lime and filtered through sand filters. The resulting sludge will be disposed in the landfill.

After filtration through activated carbon to remove organic contaminants, the water is demineralized in a series of weak and strong cation and anion exchangers. Streams produced in the regeneration of the ion exchanges are neutralized and piped to the wastewater treatment section. Wastes from the regeneration of the cation exchange system used in treating the boiler feed water are also sent to the wastewater treatment section.

The steam generation and distribution generates solid wastes in the form of bottom ash and fly ash. Both wastes are disposed of in the landfill area.

The flue gas desulfurization uses a Wellman-Lord unit to clean the exhaust gases from the coal-fired boilers. This section generates three streams: the cleaned flue gas, a sulfur dioxide–rich stream which is sent to the sulfur recovery unit, and a waste stream, containing sulfur compounds to the landfill site.

MAJOR GASIFICATION PROCESSES 207

Wastewater Treatment

This section collects and treats all the plant's water effluents, including the coal piles runoff. Water-soluble salts are concentrated by evaporation and chemically fixed by encapsulation. Slag and ash suspended in the water are removed by filtration. These solids are then disposed of in the landfill area. Only one emission, water vapor, will be discharged to the atmosphere from this section. No environmental descriptions are associated with this effluent.

Slag Handling and Disposal

This section receives all the solid wastes generated in the plant and disposes of them in a landfill area adjacent to the plant. Trace element leachate will be suppressed by locating the dump area over impermeable grounds. How long the waste material will be contained in the pond depends on several factors: the amount of rain falling in the area, the type of clay the ground is lined with, and the type of inorganic and organic material to be dumped at the site. Theoretical predictions about the rate of leachate and the rate of advance through the lining material are very unreliable due to the nature of the phenomena. The only way to forecast those rates accurately is by actual observation in the landfill.

Design of landfills is based on an assumed worst case, usually the most destructive event of the kind considered relevant to the operation of the landfill in 100 years, in this case the greatest amount of water from a given rain. Although the method is a proven one, the potential exists for a basin overfill causing the contaminated drainage to spill over.

Cooling Water System

This section has only one effluent discharging to the atmosphere. The rate of water vapor emission is such that it could create potentially hazardous conditions by producing fog across roads and highways near the plant. This situation can be avoided, however, by locating the towers in such a way that the prevailing winds will carry the water vapor away from the sensitive areas. Leakages from heat exchangers can cause pollution by evaporation from the cooling towers of the leaked material. This problem in the demonstration plant will be greatly alleviated by continuous monitoring of the cooling water systems to identify and isolate for repair the damaged equipment.

208 GASIFICATION

Plant and Instrument Air Systems

There are no effluents from this section.

Flare and Incinerator Facilities

The flare section disposes of combustible vapors produced during startup and shutdown of the plant. Emissions from this section include carbon dioxide, nitrogen, oxygen, sulfur dioxide and small amounts of unburned vapors.

The incinerator is utilized to render harmless the offgas streams coming from the Rectisol and lock-hoppers. These streams are composed mainly of carbon dioxide with small and varying amounts of hydrogen, carbon monoxide, carbonyl sulfide and some light hydrocarbons. The incinerated gases are released to the atmosphere through a 100-foot-tall stack. Sulfur dioxide emissions are not expected to exceed allowable levels, with nitrogen oxides production very low, if at all.

Tankage and Fuel Oil System

Little or no hydrocarbon emissions are expected from this section due to the low vapor pressure of the liquid by-products. Emissions from the solvents used in the operation of the plant are kept at a minimum by the use of floating roof type tanks.

Shipping and Receiving Facilities

No emissions are released from this section.

Assessment

The 300-ton/day pilot plant has operated over a 3-year period. The processes have been developed by Conoco, the British Gas Corporation and Lurgi. The developers are large organizations with extensive experience in the field.

The final pilot plant configuration operated for almost 1500 hours using coke and 3 types of coal. Eastern U.S. caking coals were successfully handled and all plant operating problems were successfully resolved.

The commercial demonstration plant utilizes only two new components: the gasifier and the methanator.

MAJOR GASIFICATION PROCESSES

The gasifier to be used in the commercial demonstration plant represents a scale-up factor of 2.6 in throughput and 100 psia (6.7 atm) in pressure over the pilot plant unit.

The Conoco-developed methanator represents a new commercial design, but gave satisfactory performance in the pilot plant.

Within the next year the slagging Lurgi process is expected to receive a contract from DOE to begin detailed engineering design of a 4000-ton/day (3600-metric ton/day) demonstration plant using commercial size components.

The design is expected to consist of only 2 gasifiers instead of the 12 used for the commercial plant described in this report. Confirmation of the suitability of the slagging Lurgi process for commercial usage will be sought as the design details become available.

TEXACO COAL GASIFICATION PROCESS

Process and Manufacture

The Texaco Coal Gasification Process manufactures either low- or medium-Btu synthesis gas in a manner that closely parallels the production of synthesis gas from petroleum fuel residues with the Texaco synthesis gas generation process. Since there are more than 75 petroleum residue plants operating in 22 countries, a large amount of operating experience with the process exists. The technical feasibility of the Texaco Coal Gasification process has been demonstrated on a 15-ton/day (14-metric ton/day) pilot plant located at Montebello, CA.

The gasifier, Figure 6-18, is a cylindrical pressure vessel, the upper section of which has been refractory lined and serves as the gasification section. The lower part extends into a water reservoir and has a restricted orifice through which slag from the gasification section will exit together with the raw gas. Immediately below the lower section of the gasifier, the water reservoir acts as a slag quencher, gas washer and gas seal to prevent pressure drops due to gas leaks in the slag removal section. The gas exit ports are located below the lower edge of the refractory lined section. This location prevents the raw gas from coming into contact with the oxygen in the feed.

Gasification of the coal occurs in a flame type environment with inadequate oxygen to consume the coal. The gasification of the coal takes place according to the reaction:

$$C + O_2 \rightarrow CO_2$$

210 GASIFICATION

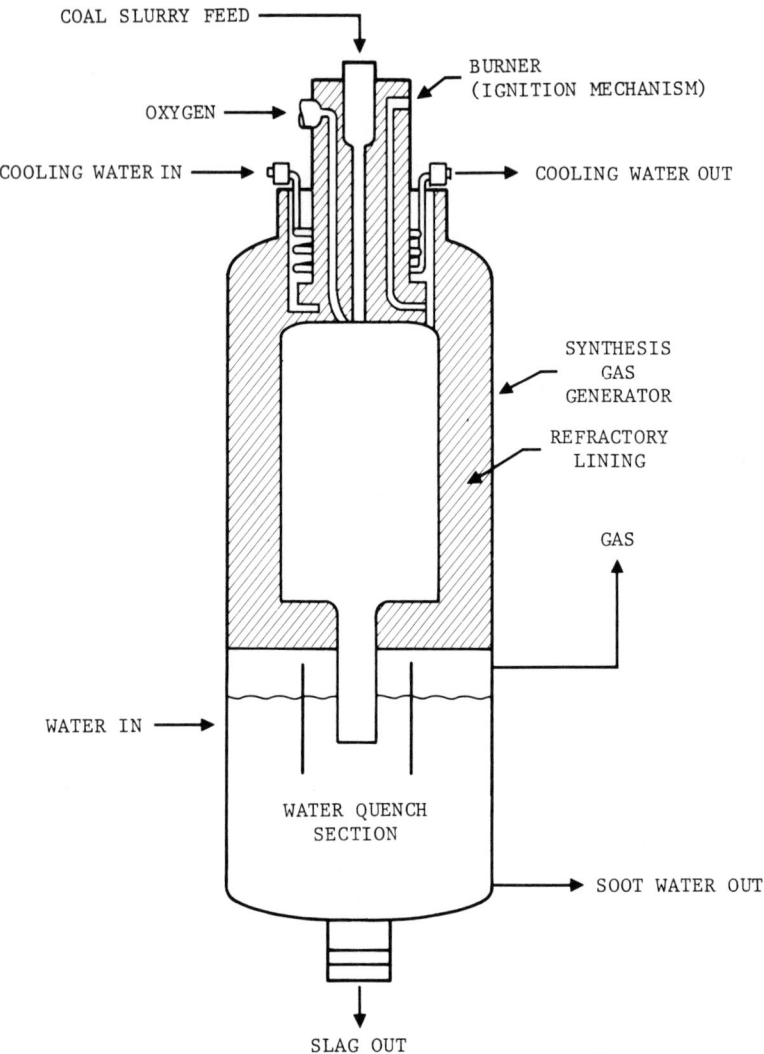

Figure 6-18. Texaco gasifier.

Heat liberated by the above reaction provides the energy needed to drive the steam/carbon and carbon/CO_2 reactions, which are endothermic.

$$C + H_2O \rightarrow CO + H_2$$
$$C + CO_2 \rightarrow 2CO$$

MAJOR GASIFICATION PROCESSES

Due to the high temperatures in the gasifier, nitrogen contained in the coal is converted to ammonia (NH_3) or free nitrogen (N_2). Most of the sulfur is converted to hydrogen sulfide (H_2S) or COS compounds.

The gasifier functions equally well with air or oxygen and does not require any additional steam in its operation because the coal is injected as a coal-water slurry. If the temperature in the reactor is above what is desirable, a moderator (usually steam) is injected to control the temperature.

In this process, coal is ground to a carefully controlled size, less than 1.5 in. (3.8 cm), and mixed with water. The slurry is then fed to the gasifier through a specially designed nozzle. The use of water to feed the coal to the gasifier solves the problem of feeding the coal continuously to the pressurized gasifier, but introduces a thermal penalty in its operation, since some of the heat of combustion from the reaction of the coal with the oxidant must be used to convert the water to steam. The penalty increases with increasing amounts of water in excess of the stoichometric amount needed for gasification.

The gasifier operates in the temperature range of 2300–2600°F (1260–1430°C) and at a pressure of about 600 psig (41 atm). Feed residence time is a few seconds, with combustion reactions proceeding almost to completion. Table 6-40 shows the operating parameters of the gasifier when operating in the air and oxygen mode. For comparison, the same data are presented for the pilot-plant gasifier. It should be noted that there has been very little pilot-plant activity in the air mode, and that the coal residence time in this mode is still under active research to obtain the optimum value. Data from the pilot plant operations have been used by Texaco Montebello Research Laboratory and by Fluor Engineers and Constructors, Inc. in the conceptual design of a combined cycle plant, which would operate on 10,000 tons/day (9091 metric ton/day) of Illinois No. 6 coal to produce 1100–1200 MW.

The composition of Illinois No. 6 coal used as feed is given in Table 6-41. The gasifier products when processing Illinois No. 6 are summarized in Table 6-42. For comparison, results of operation at the pilot plant are also shown. From the table, it can be seen that there is no production of tars or phenols. Synthesis gas (CO, H_2) yield is about 80%. The high heating value for the gas is about 320 Btu/scf (11.9 MJ/normal m^3).

Pilot-Plant Operations

The Texaco Coal Gasification Process has been developed at the Texaco Montebello Research Laboratories in California. Pilot-plant opera-

212 GASIFICATION

Table 6-40. Gasifier Operating Parameters

	Pilot Plant	Combined Cycle Plant[a]	
	Oxygen	Oxygen	Air
Cold Gas Efficiency (%)	66–73	75	68
Gas Production Rate [scf/lb (10^5 cm^3/kg)]	26–36 (16–22)	34.3 (21)	30.5
Offgas Temperature [°F (°C)]	400–500 (200–260)		
Gasification Temperature [°F (°C)]	200–2500 (1100–1370)	2360 (1300)	2300 (1260)
Gasification Pressure [psig (atm)]	350–2000 (24–140)	600 (20)	815 (55)
Residence Time (min)	<1	<1	
Percent Coal in Slurry	48–66	60	
Throughput [lb/hr-ft^2 (kg/hr-cm^2)]	300 (0.15)		
Steam/coal Ratio (wt/wt)	0.24–0.43		
Oxygen/coal Ratio (wt/wt)	0.98–1.00	0.953	1.03

[a] Estimated by Texaco, Inc.

MAJOR GASIFICATION PROCESSES 213

Table 6-41. Coal Analysis (Illinois No. 6)

Proximate Analysis (wt %)	
Moisture	4.2
Ash	9.6
Fixed Carbon	52.0
Volatile Matter	34.2
Total	100.0
Ultimate Analysis[a] (wt %)	
Carbon	77.26
Hydrogen	5.92
Oxygen	11.14
Nitrogen	1.39
Sulfur	4.29
Total	100.00
Heating Value—as Received	
Higher Heating Value (Btu/lb)	12,235
Net Heating Value (Btu/lb)	11,709

[a] Dry, ash-free basis.

Table 6-42. Gasifier Performance (Vol %)

	Combined Cycle Plant		Pilot Plant
Component	Air	Oxygen	Oxygen
CH_4	0.10	0.10	0.03
H_2	11.61	35.07	35.78
CO	19.46	51.62	44.62
CO_2	7.68	10.72	17.97
H_2S	0.52	1.22	1.02
COS	0.05	0.07	0.05
N_2	59.74	0.80	6.48[a]
Ar	0.71	0.15	
NH_3	0.11	0.24	

[a] N_2 and Ar.

tions originally were directed at developing the Texaco synthesis gas generation process in a gasifier originally developed for the partial oxidation of natural gas. Later developments led to the use of petroleum refinery residues and other solids containing carbon as feedstock. In the late 1960s solid gasification evolved to its final form. The energy crisis brought on by the 1973 Arab oil embargo greatly accelerated the development of the gasification process based on coal.

Early research into the process was carried out on a plant as shown in Figure 6-19. Coal was ground, mixed with water to form a slurry,

214 GASIFICATION

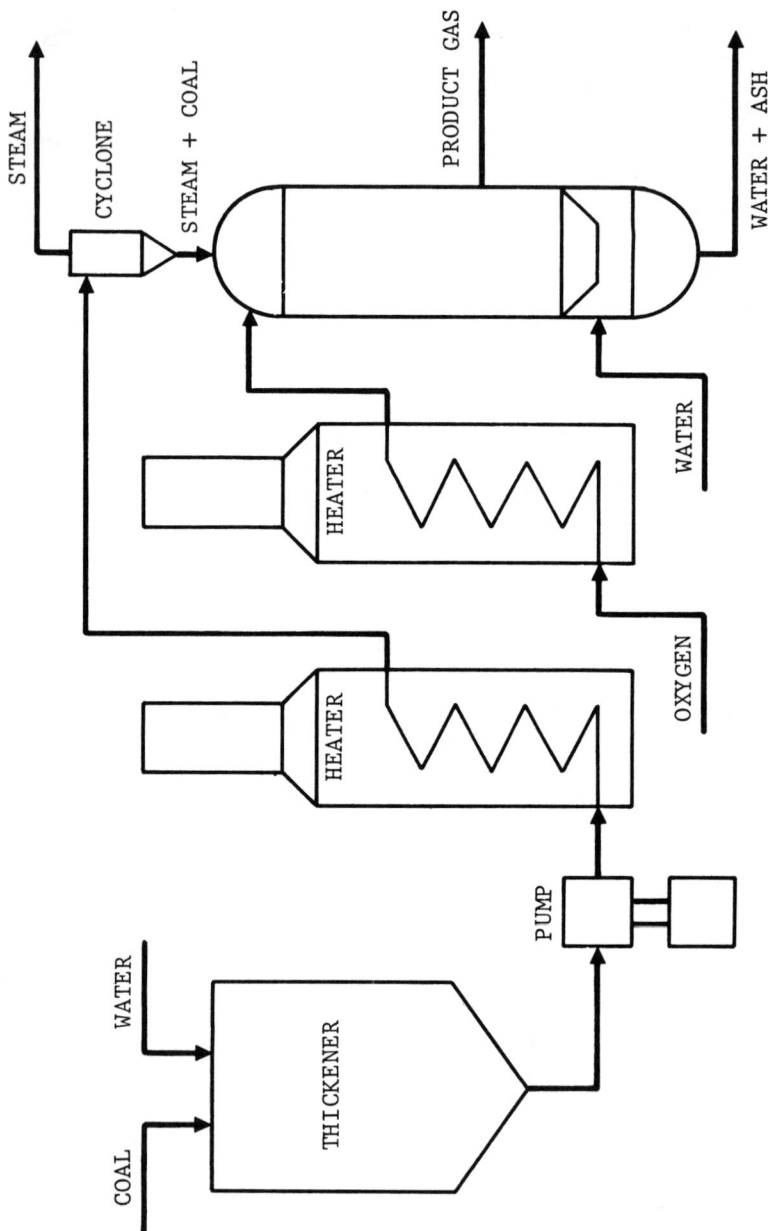

Figure 6-19. Simplified flowsheet of Texaco process at Montebello, CA, during the 1950s.

MAJOR GASIFICATION PROCESSES 215

and preheated to 700–1000°F. The actual preheat temperature depended on the type of coal. Since the amount of water needed to slurry and pump the coal was in excess of that required for gasification, the mixture was passed through a cyclone to remove about 75% of the steam. The feed then entered the gasifier either axially at the top or tangentially near the top through a water-cooled burner nozzle. Oxygen was preheated in a separate coil and fed to the reactor through a separate water cooled burner near the steam-coal nozzle. Great care had to be taken to ensure that the steam would not strike the walls of the gasifier. Molten ash flowed out of the gasifier through a restricted opening at the bottom of the gasifier into a pool of cold water. Table 6-43 shows typical data for the gasifier used then when operating with several feedstocks. Figure 6-20 shows a schematic diagram of the pilot plant as it was during the 1950s.

More recently, the process has been modified to accept the water coal slurry without preheat and with the required coal/water ratios needed for gasification, eliminating the need for the cyclone separator and the preheaters. The slurry nozzle in the modern version introduces the mixture axially to the body of the gasifier. This arrangement results in the longest possible coal-oxygen contact time and flame length, resulting in reduced feed residence time. Also, the arrangement neatly solves the problem of the oxygen contacting the walls of the gasifier. A schematic flow diagram of the pilot plant at Montebello is shown in Figure 6-21.

The process was further developed in a semicommercial-scale airblown gasifier operated at the Olin Mathieson Chemical Plant in Morgantown, WV, from mid-1956 to sometime in 1958. Details of this operation are not available.

Other activities undertaken by the synthesis gas generation research program involve the application of the gasification process to combined cycle generation and the feasibility of feeding coal liquefaction residues to the gasifier.

The objectives of the combined cycle program were:

1. to develop combustor design information for application in an industrial gas turbine using low-Btu gas as fuel; and
2. to demonstrate the feasibility of producing low-Btu fuel at the site of a power plant, as an integral part of the facility.

The program was jointly carried out by Texaco, Inc., Turbo Power and Marine Systems, Inc., a subsidiary of United Aircraft Corp., and Pratt and Whitney Aircraft, Division of United Aircraft Corporation.

Table 6-43. Typical Results for the Texaco Gasifier at Montebello, CA, During the 1950s

Fuel Type	Natural Gas	Fuel Oil	Pittsburgh Coal	Japanese Coal
Fuel Composition (wt %)				
C	74.42	85.59	77.3	64.3
H	21.94	11.38	5.3	4.9
O	1.67	0.35	5.2	15.0
N	1.98	0.72	1.4	0.9
S		1.96	2.5	1.8
Ash			7.1	13.1
Moisture			1.2	
Fuel Rate [lb/hr (kg/hr)]	1714 (777)	429 (194)		
Oxygen/Fuel (lb/lb)	1.30	1.02	0.84	0.66
Steam/Fuel (lb/lb)		0.38	0.70	0.67
Gas Composition (mol %, dry)				
CO	38.02	47.97	46.6	45.7
CO_2	2.19	3.65	11.5	13.2
H_2	59.54	47.45	38.7	37.9
CH_4	0.10	0.26	0.7	0.9
N_2	0.15	0.22	2.0	1.7
H_2S		0.44	0.7	0.6
COS		0.22		
Gas Yield (scf(dry)/lb fuel)	58.38 (3619)	50.36 (3122)	38.31 (23.75)	31.06 (19.26)
Material per 100 scf (CO + H_2)				
Fuel [lb (kg)]	17.6 (7.9)	20.8 (9.4)	30.6 (13.8)	38.5 (17.4)
Oxygen [scf (m^3)]	271.7 (7.7)	252.3 (7.1)	306 (8.6)	302 (8.5)
Steam [lb (kg)]		7.9 (3.6)	21.5 (9.8)	25.8 (11.7)
Carbon Conversion (%)	100	96.6[a]	92.1[a]	91.3[a]

[a] Calculated from reported data (cm^3/g).

MAJOR GASIFICATION PROCESSES 217

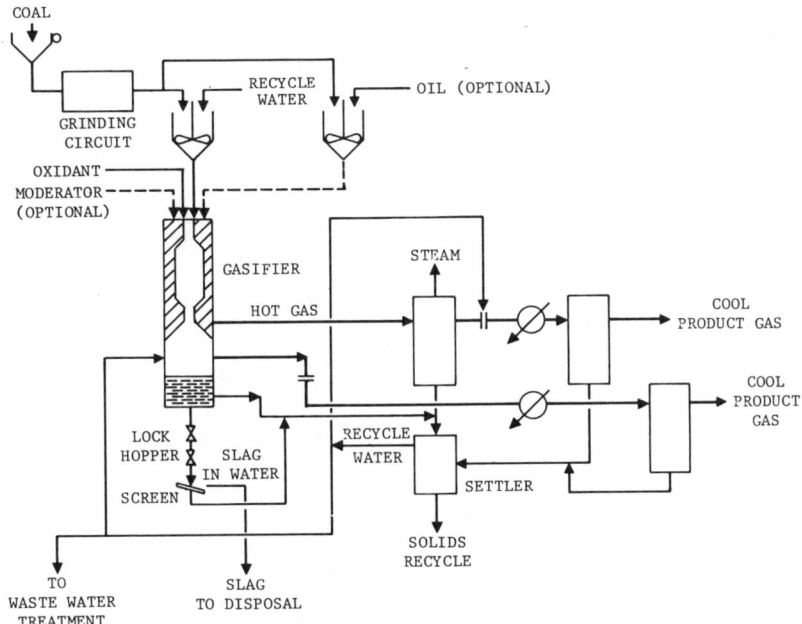

Figure 6-20. Pilot-plant flowsheet.

The combined cycle tests were conducted during 1973. The equipment utilized in the program consisted of the synthesis gas generator process and a Pratt & Whitney FT4 combustor test stand. Hydrocarbon fuel was partially oxidized in a refractory-lined pressure vessel, in a noncatalytic reaction with air or oxygen at pressures of 30–3000 psig (2–200 atm) and temperatures in the range of 1800–3000°F (900–1600°C). The resulting gas is purified of acid gas and the particulate removed prior to entering the turbines.

A 1/8 segment PWA FT4 combustor rig was used in the tests. Low-pressure air at less than 40 in. Hg (1.3 atm) was supplied to the rig by a Spencer blower driven by a 200-hp electric motor. An inline preheater fired with natural gas was used to provide heated air at engine design inlet air temperatures.

The variables and the ranges at which they were evaluated are shown in Table 6-44. Burner efficiency was found to be relatively insensitive to all variables. Figure 6-22 shows the effect of fuel heating value on efficiency. Most efficiencies fall in the 95–102% range; the fact that efficiencies exceed 100% is likely due to measurement errors in burner temperature rise.

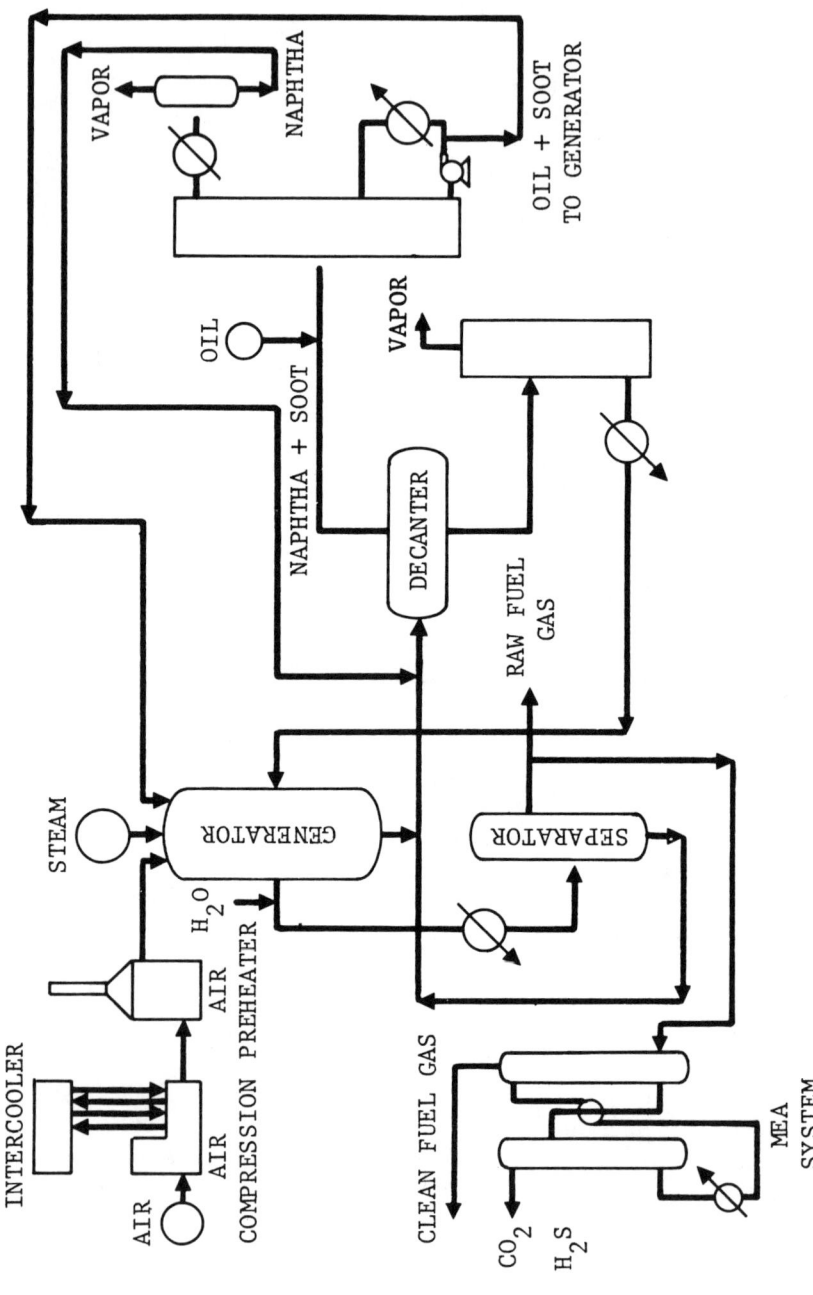

Figure 6-21. Air gasification for fuel gas manufacture (Texaco, Inc. Montebello Research Laboratory).

MAJOR GASIFICATION PROCESSES 219

Table 6-44. Combustion Chamber Parameters

Parameters	Range
Fuel-injection Velocity [ft/sec (m/sec)]	150–400 (4.6–12.2)
Burner Pressure Loss (%)	1.0–3.1
Burner Reference Velocity [f/sec (cm/sec)]	75–150 (2.3–4.6)
Injector Geometry	8 configurations
Fuel Heating Value [Btu/scf (J/cm^3)]	100–150 (3.6–5.4)

Emissions monitored during the program were nitrogen oxides (NO_x), sulfur dioxide (SO_2) and carbon monoxide (CO). Carbon monoxide emissions were expected to be higher than when natural gas is used. Test data shown in Figure 6-23 show this not to be the case. The rate of conversion of CO and CO_2 is enhanced at high pressures and a significant reduction of CO emissions is expected in combined cycle applications.

Under contract with the DOE, Texaco has instituted a program in which residues of coal liquefaction are gasified using the Texaco synthesis gas generation process. The program consists of three types of evaluations as shown in Table 6-45. Type I evaluation will determine the chemical composition and the physical properties of each test feedstock. On the basis of this evaluation the feed mode will be selected and yields and product gas composition will be estimated.

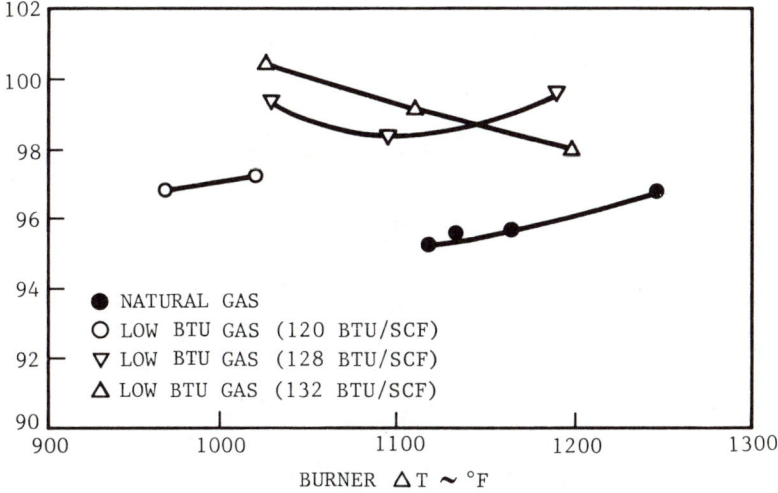

Figure 6-22. Burner efficiency vs burner ΔT.

220 GASIFICATION

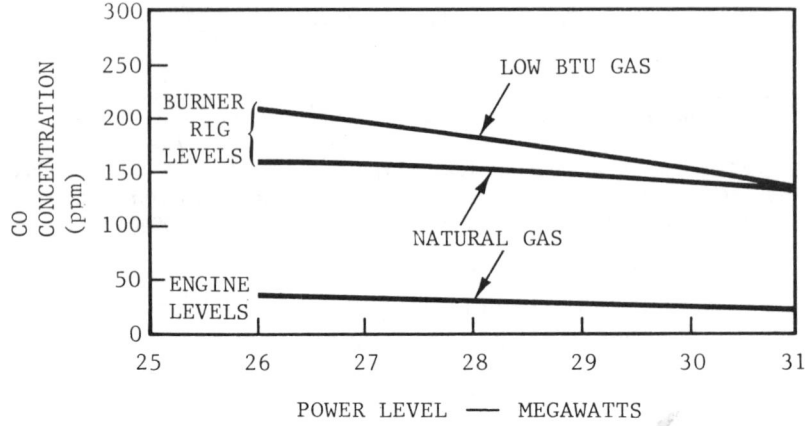

Figure 6-23. CO exhaust emission comparison. Burner rig data vs engine data.

Table 6-45. Liquefaction Residue Evaluation Program

Test Type	Description	Max. Sample Size	Min. No. of Samples	Max. No. of Samples
I	Laboratory evaluation	20 lb (9 kg)	4	15
II	Preliminary pilot plant evaluation	20 lb (9 kg)	0	4
III	Extended pilot plant evaluation	200 lb (90 kg)	0	4

In the type II evaluation pilot plant, runs will be performed with approximately 20 barrels of each material. The purpose of this evaluation is to determine the operability of the process and to allow refinements to be made in the estimates made during the type I evaluation. Table 6-46 shows a summary of the residues that have undergone types I and II evaluations. Table 6-47 shows a typical gasification result.

The type III evaluation involves extended pilot plant runs in which about 200 barrels of each candidate feedstock will be gasified.

Technical Evaluation

Historical Development

The Texaco gasification process began development in the 1940s on natural gas and has been in operation since 1946. The feedstock has been expanded to include a large number of high-sulfur petroleum waste

Table 6-46. Summary of Coal Liquefaction Residue Evaluations
Completed July 1977–June 1978

Type of Evaluation	Coal Liquefaction Process	Charge Coal	Residue	Suitability of Residue as Feedstock to Texaco Gasification Processes
I	SCR-II	Kentucky No. 9/14	Vacuum flash drum bottoms	Suitable
I	Exxon EDS	Illinois No. 6	Vacuum tower bottoms	Suitable for commercial plant; marginal for pilot plant
I	SRC	Kentucky No. 9/14	Ash concentrate from Kerr-McGee	Suitable as a slurry in water
II	SCR-II	Kentucky No. 9/14	Vacuum flash drum bottoms	Suitable; results closely check predictions
II	Exxon EDS	Illinois No. 6	Vacuum tower bottoms	Suitable for commercial plant; marginal for pilot plant; results closely check predictions
I	Rocketdyne-short-residence time liquefaction	Kentucky No. 9	Char	Marginal as a slurry in water

materials and, finally, to coal. During early years the use of coal was given low priority.

Technical details of the development are closely guarded, but one of the largest experimental coal gasifiers was built at the Morgantown, WV, plant of Olin Mathieson. It was an air-blown gasifier processing 84 ton/day (70 metric ton/day) of Pittsburgh seam coal to be used in the synthesis of ammonia and methanol. The unit was large, measuring 27 ft (9 m) high and 8 ft (2.7 m) in diameter and operated at 450 psig (30.6 atm). It is known that problems were encountered with heating of the slurry and wear in the nozzle, but data on parameters such as general plant performance, gas quality, heat losses and refractory durability are unavailable. Operations began in August 1956 and terminated in 1958.

In addition to two 15- to 20-ton/day (13.5- to 18-metric-ton/day) pilot plants using coal in Montebello, California, a 150-ton/day (135-

Table 6-47. Pilot-Plant Data from Gasification of
H-Coal Syncrude Vacuum Tower Bottoms

Product Syngas Composition[a] (mol %, dry)	
H_2	37.91
CO	55.29
CO_2	0.38
N_2	0.39
H_2S	0.06
COS	0.06
CH_4	0.14
Ar	0.03
Carbon on Coarse Slag (wt %)	0.5
Carbon on Fine Slag (wt %)	7.4
Carbon on Char (wt %)	14.1
Carbon Conversion (%)	96.9
Gasifier Pressure [psig (atm)]	350 (24)
Dry Product Gas [sch/hr (m^3/hr)]	32,900 (931)
Hydrogen Plus Carbon Monoxide [scf/hr (m^3/hr)]	30,700 (869)
Run Length (hr)	8.75

[a]All major elements plus ash are forced to yield 100% recoveries.

metric-ton/day) plant has operated for a year and a half in West Germany. The Tennessee Valley Authority will complete construction of an ammonia plant in July 1980 using a 180-ton/day gasifier. Because of the unique properties of the Texaco coal gasifier when integrated into a combined cycle gas turbine/steam turbine electric generating plant, a 90-MW(e) facility is now being designed for Coolwater, CA. Before the gasifier output is connected to the gas turbine, it will be used for one year to supply fuel to an existing boiler. The total plant capacity of 1000 ton/day (900 metric ton/day) will be supplied by seven 150-ton/day (136-metric-ton/day) gasifiers.

Technical Analysis

Because of Texaco's strict policy of not revealing technical details, it is difficult to give a definitive technical analysis of the Texaco gasifier. Because 75 plants have been operating on petroleum residuum in 33 countries since 1955, a large amount of operating experience has been gained in the production of ammonia, methanol and oxychemicals. It continues to be considered for large plants.

Technically, the process has several features that give it distinctive characteristics. The ground coal and water slurry feed is unique, although this imposes a thermal penalty due to the excess water and produces a gas having a high ratio of hydrogen to CO. It handles almost all grades

MAJOR GASIFICATION PROCESSES 223

of coal without caking problems, as well as other feedstocks. Low-rank (high-moisture) coals are not suitable because they do not form a good slurry. The high pressure operation results in a smaller, more economical facility.

The output gas is produced at the high temperature of 2400–2700°F (1300–1476°C), which requires a large heat recovery boiler and a suitable use for the steam. The high temperature operation eliminates the production of environmentally objectionable organics and simplifies the operation of the unit. More experience has been accumulated with oxygen blown units than with those using air. It should be noted that Texaco does not provide a complete plant with all required waste heat, environmental control, gas cleanup and sulfur recovery facilities. A license to use the gasifier must be accompanied by the services of an architect/engineer to design a complete facility.

Environmental Analysis

Because of the high temperatures used in the Texaco process, no tars, oils or phenols are formed. The presence of polynuclear aromatic and other hydrocarbons that pose cancer problems to workers has been detected only at minute concentrations. The glassy slag produced by the process is inert and can be safely disposed by landfilling. The Texaco gasification process achieves greater than 90% carbon conversion. The unreacted carbon floats in the slag quench water and is recycled through the reactor in the feed coal slurry. The process wastewater consists of a blowdown stream that reduces the buildup of dissolved solids. The blowdown stream can be treated by conventional water treatment technology for its content of dissolved ammonia, sulfide, formate and inorganics.

Assessment

The Texaco coal gasification process is derived from a petroleum residual process that is in widespread commercial usage. Existing plants using coal as a feedstock include a 15-ton/day (13.6-metric-ton/day) pilot plant and a 150-ton/day (36-metric-ton/day) commercial plant at Oberhausen-Holten, West Germany. Additional plants of the same size are in the planning stage.

Technical details concerning the Texaco process are considered proprietary by the developer and are not available. Adequate results from investigations at the pilot-plant level are available to state that the process works satisfactorily for runs of several hours at the 15-ton/day pilot-plant size. Design studies indicate that large plants would use multiple gasifiers, but no indication is available concerning the maximum size that is being considered.

The process produces a high temperature and operates at high thermal efficiency. An adequate heat exchanger must be employed to recover the thermal energy. Texaco does not supply a gas purification system with the gasifier. Because of the high temperature, no tars, oils, coal fines or ash are present in the gas. Ammonia and hydrogen sulfide are the principal impurities to be removed.

The Texaco gasifier has been selected for use in the EPRI-sponsored combined cycle gas turbine/steam turbine electric power plant. Studies considering substitution of air for oxygen in the gasifier have shown no economic advantage for this application.

Subject to limitations on size, the Texaco gasifier has been demonstrated to be a satisfactory producer of medium-Btu synthesis gas for chemical feedstock and hydrogen production. A 180-ton/day (163-metric-ton/day) gasifier providing hydrogen for an ammonia synthesis plant at Muscle Shoals, AL, will start up in August 1980. Tennessee Eastman will use the Texaco gasifier for methanol synthesis.

COMBUSTION ENGINEERING

Description and Schematic

The Combustion Engineering (CE) entrained bed atmospheric pressure coal gasification process began from studies conducted in the early 1970s. A gasifier for power plant applications was determined to have the optimum cost, maintenance and operational advantages when air was blown and operated at atmospheric pressure using construction methods and firing techniques similar to a pulverized coal boiler. In 1974 the design of a 5-ton/hr (4.5-metric-ton/hr) pilot plant was begun. Construction was completed in October 1977, and after a startup period, gas was produced in June 1978. The pilot plant contains coal handling, gas cleanup and char recycle facilities and is a complete but smaller version of the next phase in the development of the process, a demonstration plant supporting a 150-MW(e) utility boiler.

The CE entrained bed gasifier uses a two stage combustor. Coal is pulverized and about ⅓ is injected into a combustor where it is burned with preheated air producing a gas temperature of 3200°F

(1750°C). The hot gas rises into a reduction zone where additional coal is added with little additional air. The ⅔ of the coal added in the reduction zone gasifies with the endothermic reducing chemical reactions, dropping the temperature to 1700°F (930°C) and exits in a stream with char and H_2S. Char separated from the gas stream is recycled to the combustor, giving the process a high carbon utilization efficiency. The coal ash is melted into slag and removed from the bottom of the combustor. The product gas is cooled in a heat exchanger to 300°F (147°C), the char is removed for reinjection into the gasifier, and the product gas is spray-washed to remove particulates and desulfured using the Stretford process. In the pilot plant the clean gas is burned to preheat the incoming air. In a utility plant the gas would be reheated and burned in any of several steam or combined cycles to produce electricity.

A flow chart diagram of the CE coal gasification process is given in Figure 6-24. Coal from receiving and storage is pulverized to 200 mesh following standard utility practice. One third of the coal is pneumatically injected into the combustor, producing the heat required to drive the gasification process. The remaining coal is injected into the reducing zone to complete the gasification reaction. The hot gas and char flowing out of the reducing zone is cooled in a steam generator that is constructed using normal utility boiler practice. The steam that is produced would be utilized in the steam turbine cycle of a commercial power plant; the design parameters of the boiler must therefore be integrated into the overall design of the plant that the gasifier serves. The gas is passed through a char separator and liquid spray scrubber for removal of particulates and ammonia. The gas then passes into the gas absorber section of a sulfur removal plant operating with the Stretford process.

The Stretford process consists of four steps:

1. absorption of H_2S in an alkaline solution of Na_2CO_3 and Na_2VO_3;
2. formation of elemental sulfur by an oxidation/reduction reaction with a vanadium compound;
3. regeneration of the absorbing solution; and
4. recovery of elemental sulfur.

After absorption in the sodium carbonate solution

$$Na_2CO_3 + H_2S \rightarrow NaHS + NaHCO_3$$

the products of the reaction participate in an oxidation/reduction reac-

226 GASIFICATION

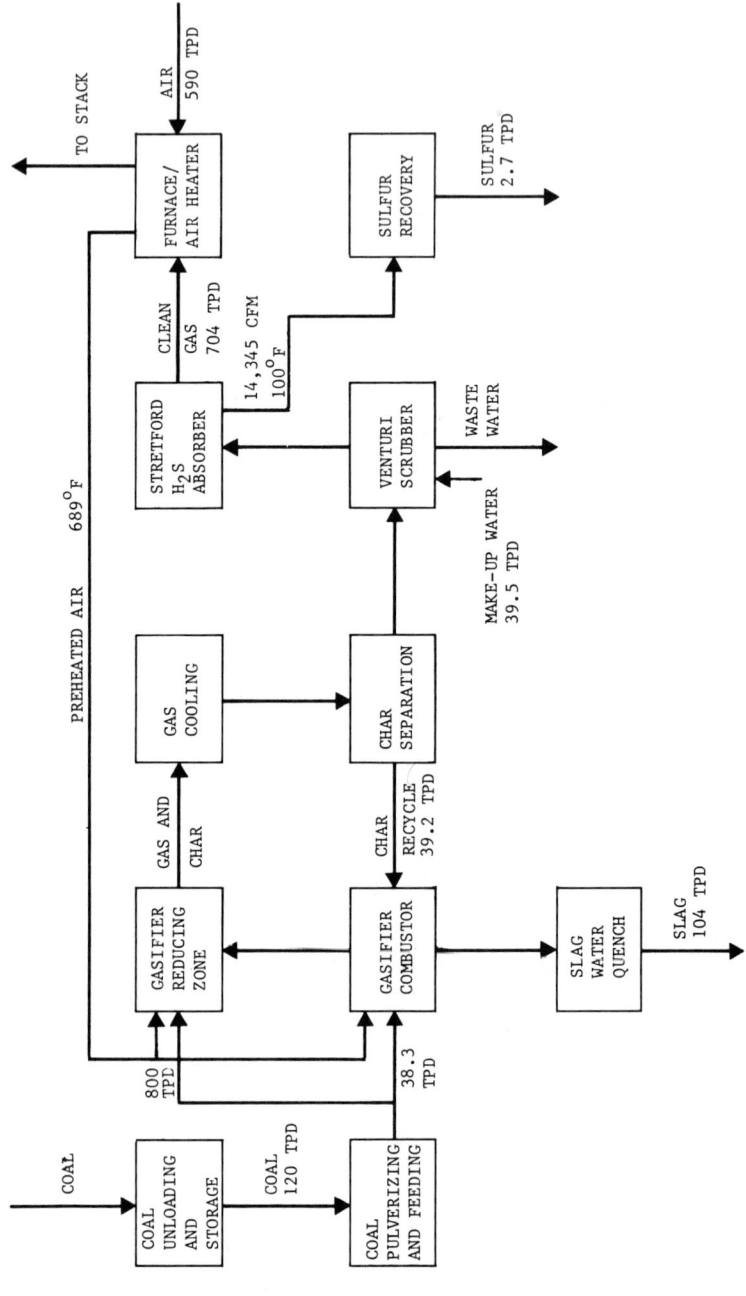

Figure 6-24. Combustion Engineering entrained-bed coal gasification process.

MAJOR GASIFICATION PROCESSES 227

tion in which sodium metavanadate and anthraquinone disulfonic acid produce a sulfur froth

$$4NaVO_3 + 2NaHS + H_2O \rightarrow Na_2V_4O_9 + 2S + 4NaOH$$

Sulfur is recovered by a rotary vacuum filter from the froth that is skimmed from the holding tank. The solution is regenerated by blowing air through a holding tank to reoxidize the vanadium into $NaVO_3$, which is then recirculated to the absorption tower. The Stretford process works well at atmospheric pressure and low CO_2 concentrations, but does not remove COS and CS_2.

After sulfur removal, the gas produced in the pilot plant is incinerated for disposal and used in preheating the gasifier air. In a utility installation the gas would be reheated using a low-temperature hot water circulation loop from the gasifier heat recovery section and burned in a steam boiler operating at atmospheric pressure using burner heads modified to utilize low-Btu gas. An alternative electric power generating cycle that has been studied is compression of the gas from atmospheric pressure to a pressure of 167–259 psig (488.6–757.8 kg/cm^2) for injection into a gas turbine used in a combined cycle generating system.

Technical Evaluation

The 120-ton/day (108-metric-ton/day) pilot plant, Figure 6-25, located at Windsor, CT, is composed of eight major processing units:

1. coal storage,
2. coal pulverization and pneumatic injection,
3. gasifier,
4. ash handling system,
5. particulate and char gas cleaning system,
6. char reinjection system,
7. Stretford gas desulfurization unit, and
8. air heater and handling system.

The processing units are scaled-down versions of equipment required for a commercial installation and constitute an integrated plant for the production and cleaning of low-Btu fuel gas. A material and flow schematic of the pilot plant is shown in Figure 6-26. The gasifier is circular in cross section, having a height of 90 ft (27 m) and a diameter of 9 ft (2.7 m). The gasifier structure itself is about 150 ft

228 GASIFICATION

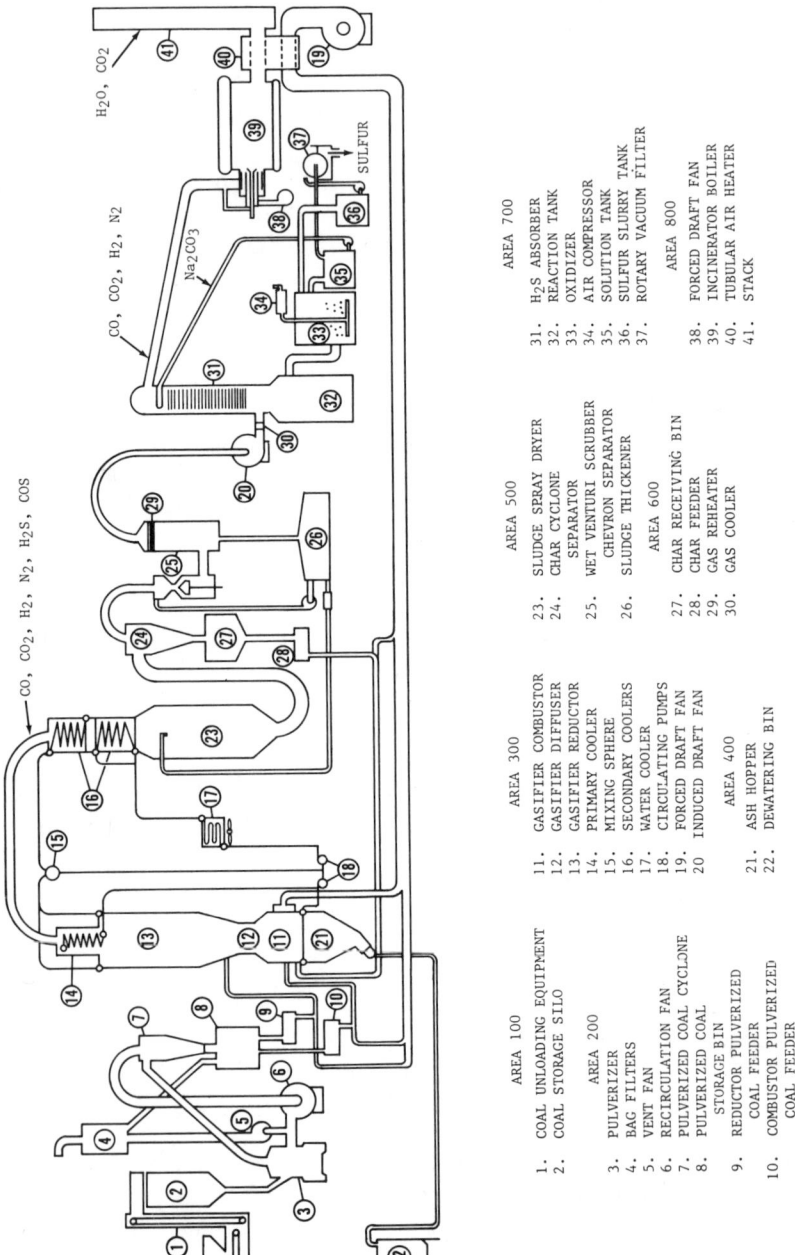

Figure 6-25. Schematic flow diagram of the CE pilot plant.

AREA 100
1. COAL UNLOADING EQUIPMENT
2. COAL STORAGE SILO

AREA 200
3. PULVERIZER
4. BAG FILTERS
5. VENT FAN
6. RECIRCULATION FAN
7. PULVERIZED COAL CYCLONE
8. PULVERIZED COAL STORAGE BIN
9. REDUCTOR PULVERIZED COAL FEEDER
10. COMBUSTOR PULVERIZED COAL FEEDER

AREA 300
11. GASIFIER COMBUSTOR
12. GASIFIER DIFFUSER
13. GASIFIER REDUCTOR
14. PRIMARY COOLER
15. MIXING SPHERE
16. SECONDARY COOLERS
17. WATER COOLER
18. CIRCULATING PUMPS
19. FORCED DRAFT FAN
20. INDUCED DRAFT FAN

AREA 400
21. ASH HOPPER
22. DEWATERING BIN

AREA 500
23. SLUDGE SPRAY DRYER
24. CHAR CYCLONE SEPARATOR
25. WET VENTURI SCRUBBER CHEVRON SEPARATOR
26. SLUDGE THICKENER

AREA 600
27. CHAR RECEIVING BIN
28. CHAR FEEDER
29. GAS REHEATER
30. GAS COOLER

AREA 700
31. H$_2$S ABSORBER
32. REACTION TANK
33. OXIDIZER
34. AIR COMPRESSOR
35. SOLUTION TANK
36. SULFUR SLURRY TANK
37. ROTARY VACUUM FILTER

AREA 800
38. FORCED DRAFT FAN
39. INCINERATOR BOILER
40. TUBULAR AIR HEATER
41. STACK

MAJOR GASIFICATION PROCESSES 229

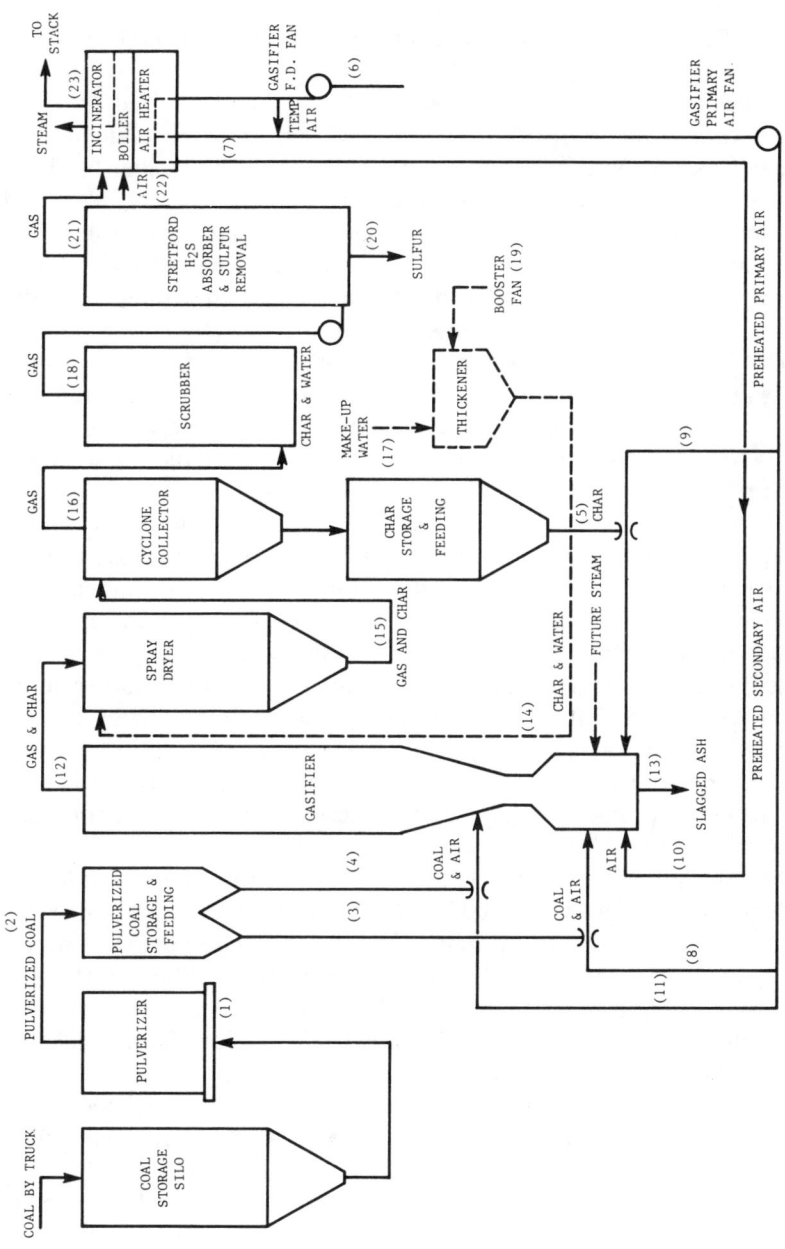

Figure 6-26. Material and flow schematic with process data for 4.5-Mg/hr pilot plant.

(45.7 m) high. The dimensions were chosen to provide the necessary gas velocity and mixing for simulating a dynamic gas flow utility sized gasifier. The CE pilot plant is the largest operating coal gasifier in the United States.

The gasifier combustor is tangentially fired with pulverized coal and recycle char. Inlet ports for air and fuel are sized to simulate the velocity and momentum in a commercial-size unit. The gasifier has been constructed with a ceramic-lined, water-cooled wall that is fabricated from welded metal strips. This construction technique is used in utility boilers and could easily be expanded to larger size. The refractory lining on the pilot unit is thicker than would be used on the commercial design, to reduce the rate of heat loss through the reduction zone walls to the rate of the commercial-size unit. A small portion of the gasifier wall is an exact duplicate of the construction to be used for the full-size unit. The test panel uses a smaller thickness of refractory lining and is heavily instrumented to measure the heat transfer data required for scale-up.

The metallurgical requirements were substantially the same in the pilot plant as in commercial practice. Carbon steel was widely used, with some usage of low-alloy ferritic steel in the gasifier and austentitic steel in the char handling system and Stretford plant. A minimum amount of corrosion has occurred. The pilot plant is heavily instrumented to obtain the data required for process verification and scale-up to commercial size. More than 500 data points are recorded by a computerized system to measure temperatures, pressures, gas, liquid and solid flowrates, gas compositions, and fuel and ash analyses, and to provide the information necessary to determine the heat and material balances.

The purpose of the pilot plant is to demonstrate the capability of the process to produce acceptable-quality gas from a variety of coals, and to provide the design information and operating experience necessary to scale the equipment up to full commercial utility boiler size. With the exception of the gasifier itself, the equipment used in the rest of the plant is commonly used in commercial practice and should pose no difficulties in a larger sized installation.

The design of the entrainment gasifier has purposefully been kept quite close to design and operation of a utility pulverized coal boiler. Operation of the unit at atmospheric pressure increases the size and capital cost of the gasifier, but minimizes the hazards and complexity of operation. Use of pulverized coal makes operation of the unit up to the point of injection into the gasifier identically the same as a coal

MAJOR GASIFICATION PROCESSES 231

fired boiler. The construction of the gasifier and heat exchanger is quite similar to a utility boiler. The char separator and gas cleanup systems are new areas, but have been selected for ease of operation. The advantages claimed for the CE entrainment gasifier are:

- easily scaled up to utility size,
- can handle almost all types of coal without special pretreatment,
- high carbon utilization (99%+) through char recycle,
- fused, water quenched ash, and
- absence of tar and oil production.

The air-blown, atmospheric-pressure, low-Btu gasifier is intended to have the reliability and performance suitable for use in a utility environment when operated by utility personnel, and is not designed for other gasifier applications such as synthetic pipeline gas production or chemical plant utilization. Some work is planned to test the CE process under oxygen blown conditions to produce a medium-Btu industrial or utility fuel.

Operating experience with the pilot plant has illustrated the types of problems that will be encountered when introducing new technologies into the utility industry. Despite a large investment of time in operator training, the initial efforts to operate the gasifier in December 1977 were troubled by numerous problems. The ability to obtain quick access to the equipment showed the advantages of an atmospheric pressure design. Many modifications to the equipment were made prior to the first successful gasification runs in June 1978. It was not until January 1979 that two long duration runs of 598 and 874 hours could be made. The material and energy balances obtained for run 7 are presented in Table 6-48. The heating value of the gas is considerably below the expected value of 112 Btu/scf (4178 J/l). This condition is expected to improve as more experience is gained in gasifier operations. Due to operating problems with the fuel feed system and particulate removal system, many of the early runs were limited to 50–70% of design capacity. At the present, more than 2000 hours of total plant operating time producing gas have been achieved. A revised test plan has been formulated to demonstrate the capability of the process. The data presented from run 7 represent the most accurate energy balance available, but do not represent plant operations at the optimum gas-making conditions. There are plans to improve the mass and energy balances further, and add overall carbon and sulfur balances to the calculations. All components of the pilot plant are now operating at design capacity.

Table 6-48. Gasifier Material and Energy Balances

	7A English	7A Metric	7B English	7B Metric	7C English	7C Metric
Inputs						
Solids Flow (lb/hr, kg/hr)						
Combustor Coal	3,285	1,491	2,280	1,035	1,870	849
Combustor Char	2,429	1,103	2,320	1,053	2,380	1,081
Reductor Coal	2,461	1,117	3,240	1,471	3,150	1,430
Air flows (lb/hr, kg/hr)						
Primary Air-Comb	11,796	5,355	10,559	4,794	10,978	4,984
Secondary Air-Comb	21,070	9,566	18,490	8,394	16,160	7,332
Primary Air-Red.	4,635	2,104	3,824	1,736	4,605	2,091
Scanner/temp. Air	2,891	1,313	4,168	1,892	3,809	1,729
Air Temperatures (°F, °C)						
Primary Air	379	193	691	366	676	358
Secondary Air	687	364	691	366	676	358
Scanner/temp. Air	66	19	68	20	76	24
Energy Streams (10^6 Btu/hr, MW)						
Combustor Coal	43.10	12.63	29.23			
Combustor Char	18.46	5.41	17.66	8.77	24.55	7.19
Reductor Coal	32.29	9.46	42.53	5.17	18.12	5.31
Combustor Air	4.48	1.31	3.96	12.46	41.36	12.12
Reductor Air	0.41	0.12	0.34	1.16	3.55	1.04
				0.10	0.41	0.12
Totals						
Material in (lb, kg)	48,567	22,049	44,881	20,375	42,942	19,496
Energy in (10^6 Btu/hr, MW)	98.74	28.93	94.42	27.66	87.99	25.78

MAJOR GASIFICATION PROCESSES 233

Outputs				
Solid Flows (lb/hr, kg/hr)				
Char in Gas	2,149	2,278	1,034	2,018
	976	594	270	693
Slag	727			
	330			
Gas Flows (lb/hr, kg/hr)				
CO	5,482	4,471	2,484	4,201
	2,489	4,929	2,238	4,874
CO_2	5,391	140	64	97
	2,448	1,575	715	1,734
H_2	133			
	60			
H_2O	1,798	27,620	12,539	28,740
	816	133	51	103
N_2	31,010	21	10	20
	14,079	44	20	43
H_2S	122			43
COS	55			
CH_4				
O_2	96			
Gas Temperature (°F, °C)	1,942	1,975	1,080	1,942
	1,061	64.40		49.6
Gas Heating Value[a] (Btu/scf, MJ/m³)	56.26	2.08	2.38	1.84
Energy Streams (10⁶ Btu/hr, MW)				
Gas Chem. Energy	32.78	33.24	9.74	25.01
Gas Sens. Energy	26.24	23.75	6.96	23.27
Gasifier Absorp.	23.15	19.66	5.76	22.06
Char	17.57	18.65	5.46	16.49
Slag	1.83	1.52	0.45	1.72
Totals				
Material out (lb, kg)	46,908	42,785	19,425	42,566
Energy out	101.57	96.82	28.37	88.55
Closure[b] (%)				
Material Balance	+3.4	+4.7	+4.7	+0.9
Energy Balance	−2.9	−2.6	−2.6	−0.6

				916
				315
				1,907
				2,213
				44
				787
				13,048
				47
				9
				20
				20
				1,061
				7.33
				6.82
				6.46
				4.83
				0.50
				19,326
				25.94

[a] Dry, and H_2S-free.
[b] (in - out)/in.

234 GASIFICATION

All runs to date have been conducted using Pittsburgh No. 8, a bituminous caking coal with the properties given in Tables 6-49 and 6-50. During 1980 gasification runs will start using four additional coals:

1. Illinois No. 6
2. Montana (Rosebud Mine)
3. Arizona (Four Corners)
4. Texas lignite

Economic Evaluation

In 1978 CE performed a study to compare the capital and operating costs of plants using the CE coal gasifier with the costs of a conventional coal firing plant using a flue gas desulfurization unit. Three 600-MW(e) electric power plants were designed incorporating the same-sized coal gasifier:

1. a conventional steam plant (Figure 6-27) using the integrated CE gasifier;
2. a combined cycle plant (Figure 6-28) with 1800°F (980°C) gas turbines and supplementary fired heat recovery boiler
3. a combined cycle plant (Figure 6-29) using an advanced gas turbine with a 2200°F (1200°C) firing temperature and unfired waste heat recovery boilers.

Table 6-49. Coal Used and Production Gas Composition

Proximate Analysis (wt %)	
Volatile matter	37.5
Fixed carbon	51.5
Moisture (as fired)	0.4
Ash	10.6
Total	100.0
Ultimate Analysis (wt %)	
Carbon	72.8
Hydrogen	4.9
Oxygen	7.9
Nitrogen	1.3
Sulfur	2.1
Moisture	0.4
Ash	10.6
Total	100
HHV [kJ/g (Btu/lb)]	30.58 (13,160)

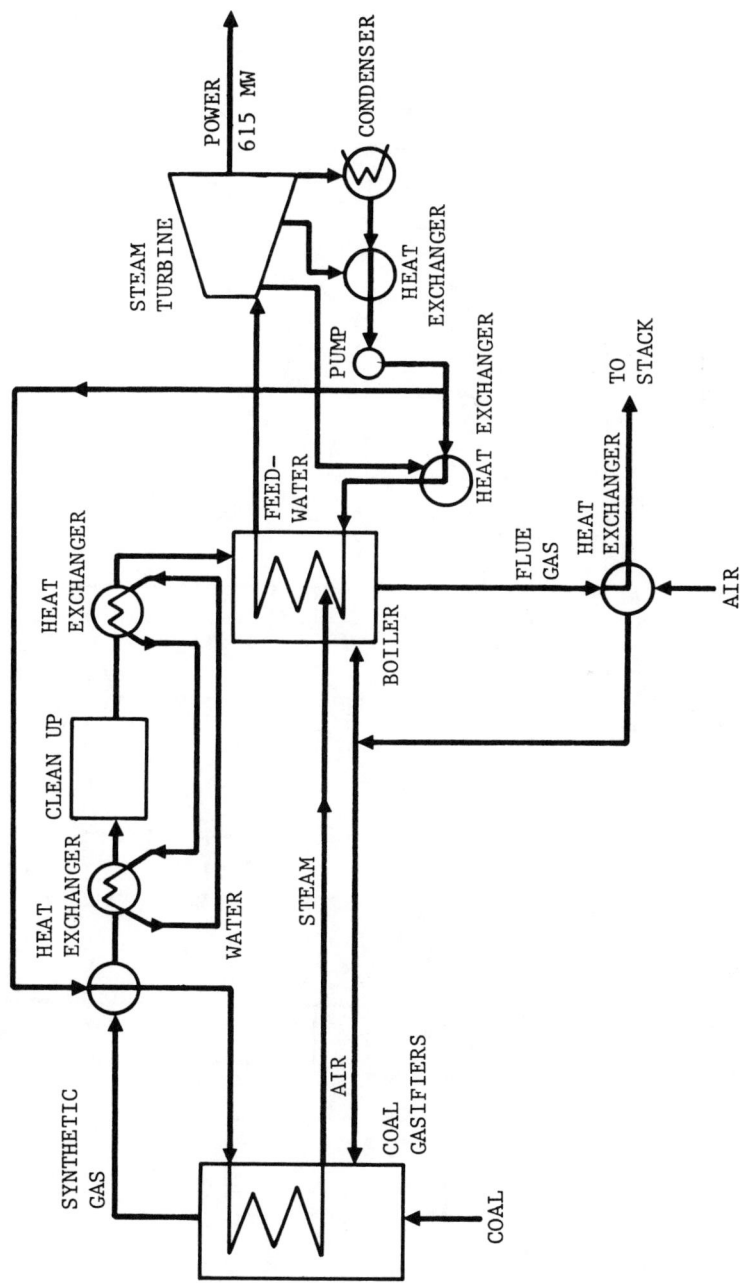

Figure 6-27. Schematic of conventional steam plant with atmospheric gasifiers.

236 GASIFICATION

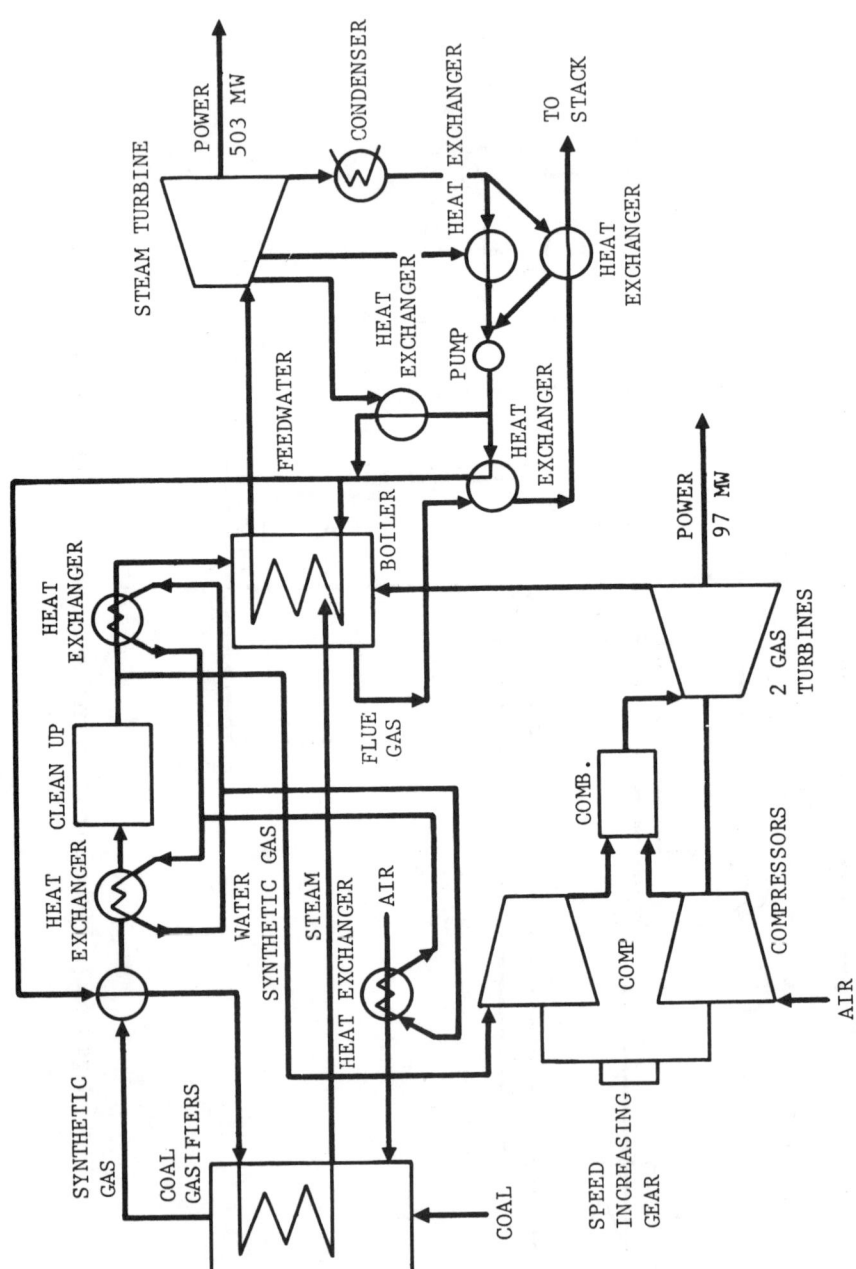

Figure 6-28. Schematic of combined cycle plant with supplementary fired boiler and atmospheric gasifiers.

MAJOR GASIFICATION PROCESSES 237

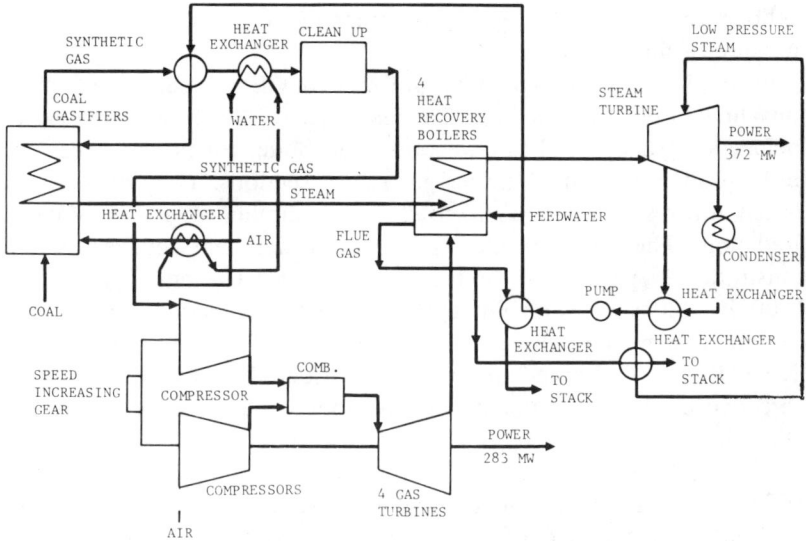

Figure 6-29. Schematic of combined cycle plant with heat recovery boilers and atmospheric gasifiers.

Table 6-50. Range of Product Gas Compositions

Components	Predicted (vol %)	Initial Results (vol %)
CO_2	4.4	12.5–3.0
CO	22.9	7.5–30.0
H_2	10.0	5.0–9.5
N_2	62.7	74.0–58.0
CH_4	0.0	0.0–3.0
Other Hydrocarbons	0.0	0.0–1.0
H_2S	100 ppm	10–55 ppm
HHV	3.9 MJ/m³	1.5–5.4 MJ/m³
	(106 Btu/scf)	(40–145 Btu/scf)

All systems would utilize a 2500-psi (7315-kg/cm²) steam cycle with 1000°F (538°C) superheat temperature and 1000°F (538°C) reheat. The heat recovered in the gasifier would be utilized in the steam cycle. The 1800°F (980°C) combined cycle plant would utilize two gas turbines and a single gas-fired heat recovery boiler. The 2200°F (1200°C) design utilizes four gas turbines, each with a separate 55-MW(e) heat recovery boiler. The higher gas turbine firing temperature results in a higher exhaust temperature that does not require supplementary firing. Each of the gas turbines mechanically drives a compressor to boost the low-Btu fuel gas from atmospheric pressure to the working pressure of the gas turbines.

Each plant was designed to utilize two CE gasifiers rated at 100 tons/hr (90 metric tons/hr). The heat recovery sections in the gasifiers include superheater, reheater, economizer, boiler for process steam used in the gasifier and liquid couple gas reheat sections. The all steam cycle design utilizes a low-Btu gas-fired boiler similar to a 435 MWe oil-fired unit. The steam produced in the gasifier heat recovery sections constitutes 27 percent of the total steam cycle. The boiler used for the 1800°F (980°C) gas turbine cycle is a smaller version of the all steam design. The higher temperature gas turbine design was provided by Brown Boveri Turbomachinery and is considered to be an easily achieved objective by 1985. Four independent unfired heat recovery boilers are utilized.

The capital costs including field erection are computed in November 1976 dollars. Estimates include engineering, construction management, contractor's fee and contingency costs. Interest during a 5-year construction period is estimated at a total of 20% of total cost. The cost of capital for 30-year financing is covered by an 18% fixed charge. Escalation of labor, materials and fuel costs for a 30-year period double the costs that were in effect during 1976. Operating, labor and materials costs vary with the design of the plants and staffing required for the steam and gas turbine portions. Operating costs are computed on an assumed 7000 hr/year operating duty at 100% load.

The results of an economic analysis of the three designs using the gasifier and a reference steam plant with a flue gas scrubber are shown in Table 6-51. With all analyses performed on a consistent basis, the predicted cost of electricity declines as the technological level of the plant becomes more sophisticated. A significant decrease in the heat rate and the cost of electricity occurs when the high firing temperature gas turbine combined cycle is used. The predicted cost of electricity is about 10% lower than the cost for a steam plant using scrubbers.

Environmental Analysis

Similar to the Texaco process, the Combustion Engineering gasifier produces no tars, oils or phenols. Although not confirmed by data from actual pilot plant testing, it is expected that the high temperatures used in gasification will minimize the concentrations of polynuclear aromatic and other hydrocarbons having adverse worker health effects, except in the char recirculation loop. An appropriate worker sanitation pro-

MAJOR GASIFICATION PROCESSES 239

Table 6-51. Operating Cost Data
(November 1976, 100% Load for 7000 hr/yr)

	Pulverized Coal with Scrubbers	Two Gasifiers with Steam Boiler	983°C Gas Turbine	1,205°C Gas Turbine
Heat rate (Btu/kWh)	9,456	9,666	9,069	8,223
Efficiency (%)	36.1	35.3	37.6	41.5
Net MW	524.9	574.6	572.0	630.8
Annual Power (10^6 kWh)	3,674	4,022	4,004	4,416
Annual Costs ($$10^3$)				
Capital @ 18% F.C.	62,419	70,115	73,110	79.901
Labor & materials[a]	21,336	14,516	15,178	14.240
Fuel @ $2.00/$10^6$ Btu[a]	69.483	77.752	72,620	72,620
Total W/$2.00 Coal	153,238	162,383	160,908	166,761
Levelized mills/kWh W/$2.00	41.7	40.4	40.2	37.8

[a]Escalated and levelized for 30-year plant life with declining load after 15 years.

gram will have to be employed to protect workers during maintenance operations. Work on the boiler tubes would be likely to produce exposure to such material. The fused slag produced by the gasifier should present minimum leaching problems in a landfill. Wastewater from the gas scrubbing process must be analyzed to determine its chemical content and the treatment processes that will be required.

Assessment

The CE gasifier has been designed to be constructed and operated similarly to conventional utility boiler practice. If the reliability of the design and ease of maintenance by utility personnel are confirmed in practice, they could be the most important factors in gasifier acceptability in utility operations.

The atmospheric pressure design makes the unit larger but simpler than pressurized designs. The output is better suited to be coupled to an atmospheric pressure boiler than to a combined cycle gas turbine requiring pressurization.

Retrofit installations of gasifiers onto existing steam boilers are presently in disfavor in the United States. The use of coal oil mixtures and the development of high-temperature combined cycles are being emphasized instead. However, one of the most immediate uses of the CE process is for existing boiler retrofit.

The low-Btu gas that is produced is better suited as a boiler fuel than for other applications, such as chemical plant or pipeline gas synthesis.

Runs using oxygen to produce medium-Btu gas are planned, but the CE gasification process has been primarily developed to produce low-Btu gas.

The CE gasifier pilot plant is currently the largest operating gasifier in the United States. Because of the initial operating problems, it has only reached its full potential since June 1979. Additional test run data will be necessary before a judgment of performance can be made. Scale-up to the several hundred megawatt size required for commercial utility scale operations can be done with more confidence in achieving successful operation than with other gasifiers.

BIBLIOGRAPHY

Allen, D. W., and W. H. Yeu. "Methanator Design and Operation," *Chem. Eng. Prog.* 69(1):75-79 (1973).
Atlantic Research Corp. "Development of Coatings for Protection of Coal During Transport and Storage," COO-4632-2, prepared for the U.S. Department of Energy (1978).
Continental Oil Co. "Phase I: The Pipeline Gas Demonstration Plant. Design and Evaluation of Commercial Plant, Vols. 1-4," prepared for the U.S. Department of Energy, FE-2542-10 (1979).
Continental Oil Co. "Phase I: The Pipeline Gas Demonstration Plant. Technical Support Program Report," FE-2542-13, prepared for the U.S. Department of Energy (1979).
Covell, R. B., and M. J. Hargrove. "Power Cycle Evaluation of the C-E Coal Gasification," paper presented at the American Power Conference, April 1979.
Dravo Corp. "Handbook of Gasifiers and Gas Treatment Systems," FE-1772-11, prepared for the U.S. Energy Research Agency (1976).
Fluor Engineers and Construction, Inc. "Economic Studies of Coal Gasification Combined Cycle Systems for Electric Power Generation," EPRI AF-642, prepared for the Electric Power Research Institute (1978).
Fluor Engineers and Construction, Inc. "Economics of Texaco Gasification —Combined Cycle Systems," EPRI AF-753, prepared for the Electric Power Research Institute (1978).
MITRE Corp. "Compilation and Evaluation of Leaching Test Methods," MTR-7758, prepared for the U.S. Environmental Protection Agency (1978).
MITRE Corp. "Environmental Data for Energy Policy Analysis. Volume I: Summary," M78-74, prepared for the U.S. Department of Energy (1978).
Patterson, R. C. "Coal Gasification for Power Plant Fuel," paper presented at the VGB Conference on Gasification of Coal in Power Engineering, March 1979.
Patterson, R. C., and S. L. Darling. "A Low Btu Coal Gasification System," paper presented at the 72nd Annual Meeting of the AICLE, November, 1979.

"R&D Status Report: British Gas Corporation Slagging Gasifier," *EPRI J.* (January/February 1979).

Savage, P. R. "Slagging Gasifier Aims for SNG Market," *Chem. Eng.* 84 (19):108-109 (1977).

Sudbury, J. D. "A Demonstration of the Slagging Gasifier," in Proceedings of the Eighth Synthetic Pipeline Gas Symposium (1978).

TRW. "Environmental Assessment Data Base for High-Btu Gasification Technology: Volume II. Appendices A, B, and C," EPA-600/7-78-186b, prepared for the U.S. Environmental Protection Agency (1978).

TRW. "Environmental Assessment Data Base for High-Btu Gasification Technology: Volume II. Appendices D, E, and F," EPA-600/7-78-186c, prepared for the U.S. Environmental Protection Agency (1978).

Wilson, H. S. "Procedures for Coal Storage at Industrial Plants," *Plant Eng.* (June 23, 1977), p. 121-1.

CHAPTER 7
MINOR GASIFICATION PROCESSES

SHELL-KOPPERS

Description and Schematic

The Shell-Koppers development program is a joint venture of Shell Internationale Research Mij. and Krupp-Koppers, aimed at the commercialization of a coal gasification process operating in the entrained-bed mode at elevated pressures and, depending on the moisture content of the coal, little or no steam. The process has been developed in a 6-ton/day (5.4-metric-ton/day) process development unit, located at Shell's Amsterdam laboratory since December 1976. A 150-ton/day (136-metric-ton/day) pilot plant scheduled to start operations in early 1978 is expected to provide the necessary operating experience to proceed with the commercialization phase of the project. The pilot plant is scheduled to operate until 1980, at which time the process is expected to be ready for commercial-scale implementation.

The gasification reactions occur in a flame-like environment in the presence of excess carbon to ensure the formation of synthesis gas. Gasification proceeds according to the reactions:

$$C + 0.5O_2 \rightarrow CO \qquad (1)$$
$$CO + H_2O \rightarrow CO_2 + H_2 \qquad (2)$$
$$CO + 3H_2 \rightarrow CH_4 + H_2O \qquad (3)$$

At the high temperatures and pressures at which the gasifier operates, Equations 1 and 2 are favored.

In this process, oxygen and finely ground and dried coal, 90% of which is smaller than 3.6×10^{-3} in. (90 μ), are fed through nozzle

244 GASIFICATION

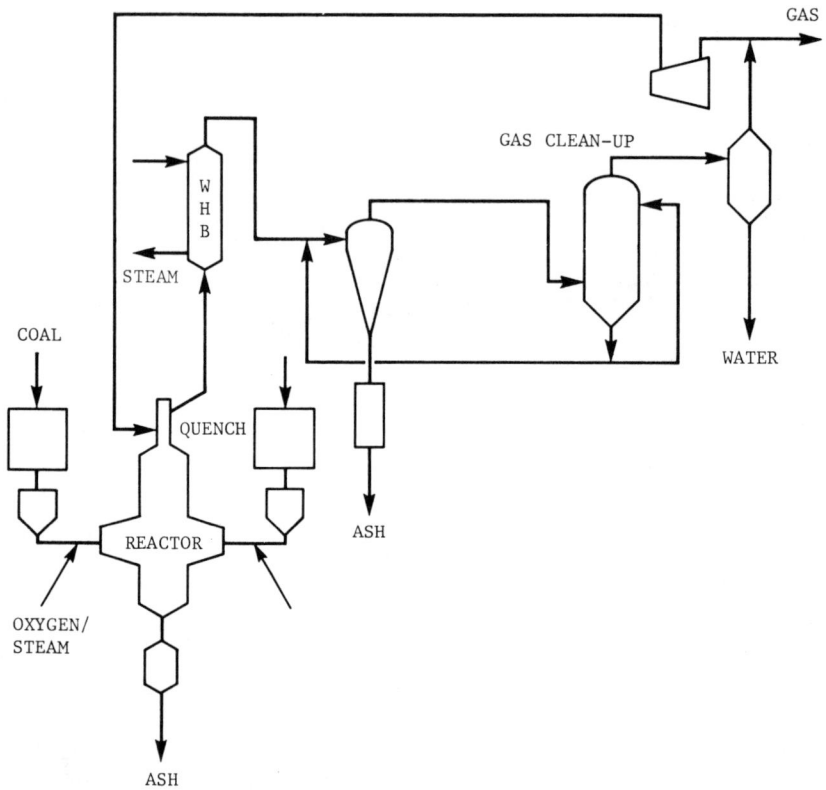

Figure 7-1. Schematic flow diagram of the Shell-Koppers process.

burners placed directly opposite each other. A schematic diagram of the process is shown in Figure 7-1. The gasifier operating parameters are shown in Table 7-1. The ash flowing out of the reactor is collected in a water-filled compartment at the bottom of the reactor.

The hot raw gases are passed through a proprietary system consisting of a cyclone and scrubbers to remove the entrained matter in the gasifier. Table 7-2 shows the composition of the raw gas resulting from the gasification of four different coals with the composition shown in Table 7-3. From the table it can be seen that the moisture content varies from 10 to 50 wt %. Steam consumption is inversely proportional to the moisture content, varying from 50% of the base case to 0%. Coal input increases as the rank of the coal decreases, as shown in Table 7-4. The efficiency of the gasifier is minimally affected by the different feedstocks, as shown in the table for the ranges of ash content investigated

MINOR GASIFICATION PROCESSES 245

Table 7-1. Shell-Koppers Gasifier Operating Conditions[a]

Temperature [°F (°C)]	>2550 (1400)
Coal Size [in. (cm)]	3.6×10^{-3} (90×10^{-4})
Oxygen Feed (ton/ton coal)	0.7–1.0
Steam, (ton/ton coal)	0.0–0.07
Gas production rate[b] [scf/lb (m³/metric ton)]	34.95 (2000)
CO/H_2 Ratio	2.0–2.4
Heating Value [Btu/scf (kcal/m³)]	303 (2700)
Thermal Efficiency (%)	77–80
Cold Gas Efficiency (%)	92–94
Offgas Temperature [°F (°C)]	<2700 (1500)

[a]Source: Kraayveld and van der Burgt, 1978.
[b]For good-quality bituminous coals.

Table 7-2. Shell-Koppers Coal Gasification Dry Synthesis Gas Composition[a]

Component (vol %)	West German Bituminous		Wyodak Lignite	Australian Brown Coal (Yallourn)
	Low-Ash	High-Ash		
H_2	31.3	30.2	30.1	28.6
CO	65.6	66.5	66.1	65.8
CO_2	1.5	1.8	2.5	4.7
CH_4	0.4	0.3	0.4	0.1
H_2S	0.4	0.4	0.2	0.1
N_2	0.6	0.6	0.5	0.5
A	0.2	0.2	0.2	0.2

[a]Source: Kraayveld and van der Burgt, 1978.

Table 7-3. Shell-Koppers Coal Gasification Coal Feed Analyses[a]

	West German Bituminous		Wyodak Lignite	Australian Brown Coal (Yallourn)
	Low-Ash	High-Ash		
Carbon	66.5	51.4	44.6	33.0
Hydrogen	4.3	3.3	3.5	2.3
Oxygen	8.0	6.2	9.9	13.1
Nitrogen	1.0	0.8	0.6	0.3
Sulfur	1.1	0.9	0.4	0.1
Ash	9.	27.4	6.0	1.2
Moisture	10.0	10.0	35.0	50.0
Heating Value (LHV)				
kcal/kg	6,300	4,860	4,100	2,680
Btu/lb	11,300	8,700	7,400	4,800

[a]Source: Kraayveld and van der Burgt, 1978.

246 GASIFICATION

Table 7-4. Shell-Koppers Coal Gasification Relative Amounts of Coal, Oxygen and Steam for Different Coals[a]

Constant Plant Capacity	West German Bituminous		Wyodak Lignite	Australian Brown Coal (Yallourn)
	Low-Ash (base case)	High-Ash		
Coal Intake	100	132	152	232
Oxygen Demand	100	102	105	125
Steam Demand	100	50	20	
Overall Thermal Efficiency (%) (LHV - Basis)	77	74	77	76

[a]Source: Kraayveld and van der Burgt, 1978.

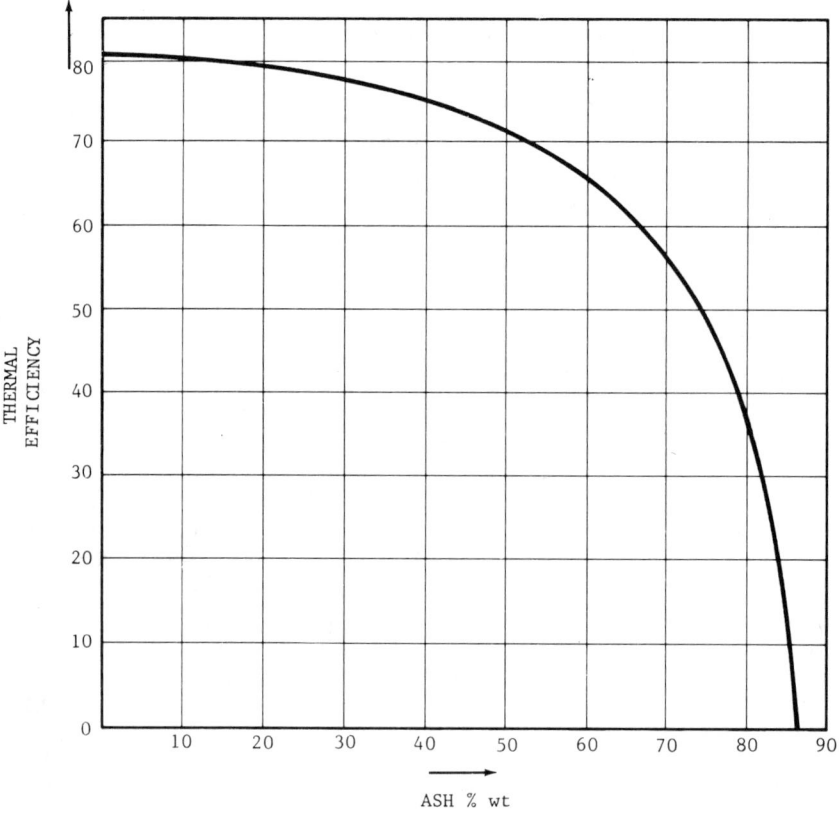

Figure 7-2. Overall thermal efficiency.

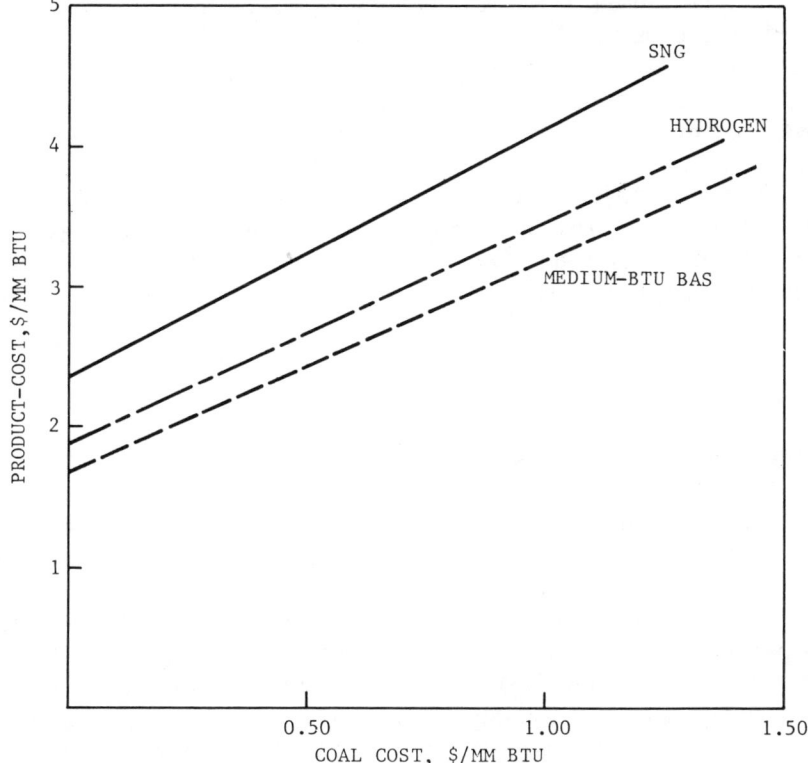

Figure 7-3. Cost of gasification product.

(10–30 wt %). The overall thermal efficiency changed as shown in Figure 7-2.

Economic Analysis

Investment required for a 141×10^6 scf (4×10^6 m³) per day synthesis gas plant is estimated at $120 million (mid-1977 dollars). Included in this figure are the processing of 10% ash, 10% moisture hard coal, and the investments for coal handling and storage, coal mill and dryer, oxygen plant, water treatment, ash disposal facilities and offsite facilities. Not included in the estimate is capital investment required for further processing of the gas.

Cost of the synthesis gas before and after conversion into SNG and hydrogen is shown in Figure 7-3 on a heating value basis as a function

248 GASIFICATION

of the coal feed unit cost for three different gases: SNG, hydrogen and medium-Btu gas.

Assessment

The Shell-Koppers method is a new process that has not been extensively developed. Two years of experience with the gasifier has been obtained with a 150-ton/day (136-metric-ton/day) installation. There are no plans for another unit.

COGAS

Description and Schematic

The COGAS process consists of three subprocesses: multistage fluidized bed pyrolysis, char gasification and processes for upgrading the raw product to synthetic crude oil and synthetic natural gas.

The technical feasibility of the processes has been demonstrated separately. The pyrolyzing process has been demonstrated in a 36-ton/day (32.7-metric-ton/day) pilot plant located in Princeton, NJ used to demonstrate the feasibility of the COED process. The gasification process has been demonstrated in a 50-ton/day (45.4-metric-ton/day) pilot plant located in Leatherhead, England. Chars from the COED pilot plant were used as feedstock for the gasification pilot plant. Data from the operation of both pilot plants (Table 7-5) have been used by the COGAS Development Corporation, of Princeton, NJ, in the conceptual design of a commercial plant based on the COGAS process.

The pyrolysis section of the process is similar in its operation to the COED process shown in Figure 7-4. The difference between the processes lies in how the heat needed for pyrolysis is generated. In the COED process, this is accomplished by reacting some of the char with oxygen to yield hot synthesis gas, which is then passed through the other stages of the process in countercurrent fashion. In the COGAS process, heat is provided from synthesis gas generated in the gasifier. The gasifier does not generate its products as a result of chemical processes. The principal mechanism at work in the pyrolysis is devolatilization, a physical process which results from the action of heat on the feedstock. Figure 7-5 shows a schematic diagram of the pyrolysis and char combustion processes.

Table 7-5. Summary of COED Process Development Unit: Results on All Coals Tested[a]

	Utah A-Seam	Illinois No. 6	Colorado Bear	New Mexico McKinley	Wyoming Rock Springs	Wyoming Glenrock	Montana Colstrip	Indiana No. 5	Indiana No. 6	W. Kentucky Paradise No. 9
Moisture as Received (wt %)	5.0	10.0	0.5	15.0	9.0	24.0	23.0	6.0	10.0	6.0
Proximate Analysis (wt % dry)										
Volatile Matter	42.9	38.6	36.5	41.8	44.6	40.3	36.2	37.0	36.8	38.9
Fixed Carbon	51.0	50.0	53.6	49.6	52.1	39.8	52.5	51.2	51.4	53.9
Ash	7.1	11.4	9.9	8.6	3.3	19.9	11.3	12.8	11.8	7.2
Htg. Value as Received										
[Btu/lb (kcal/kg)]	12,700	10,900	12,800	10,000	12,480	7,500	8,720	11,660	11,370	12,770
	7,048	6,050	6,100	7,100	6,926	4,162	4,840	6,471	6,310	7,087
Product Yields, per Ton										
Char (tons)	0.375	0.40	0.432	0.289	0.368	0.297	0.368	0.448	0.426	0.545
(metric tons)	0.340	0.362	0.392	0.262	0.334	0.270	0.334	0.406	0.386	0.494
Crude Oil (bbl)	1.37	1.04	1.40	0.93	1.14	9.425	0.373	1.13	0.98	1.06
Gas (scf, CO_2-free)	7,900	7,720	8,470	11,100	8,290	5,250	4,250	5,770	6,910	
(normal m^3)	224	219	240	314	235	149	120	163	196	

[a] Bloom, 1976.

250 GASIFICATION

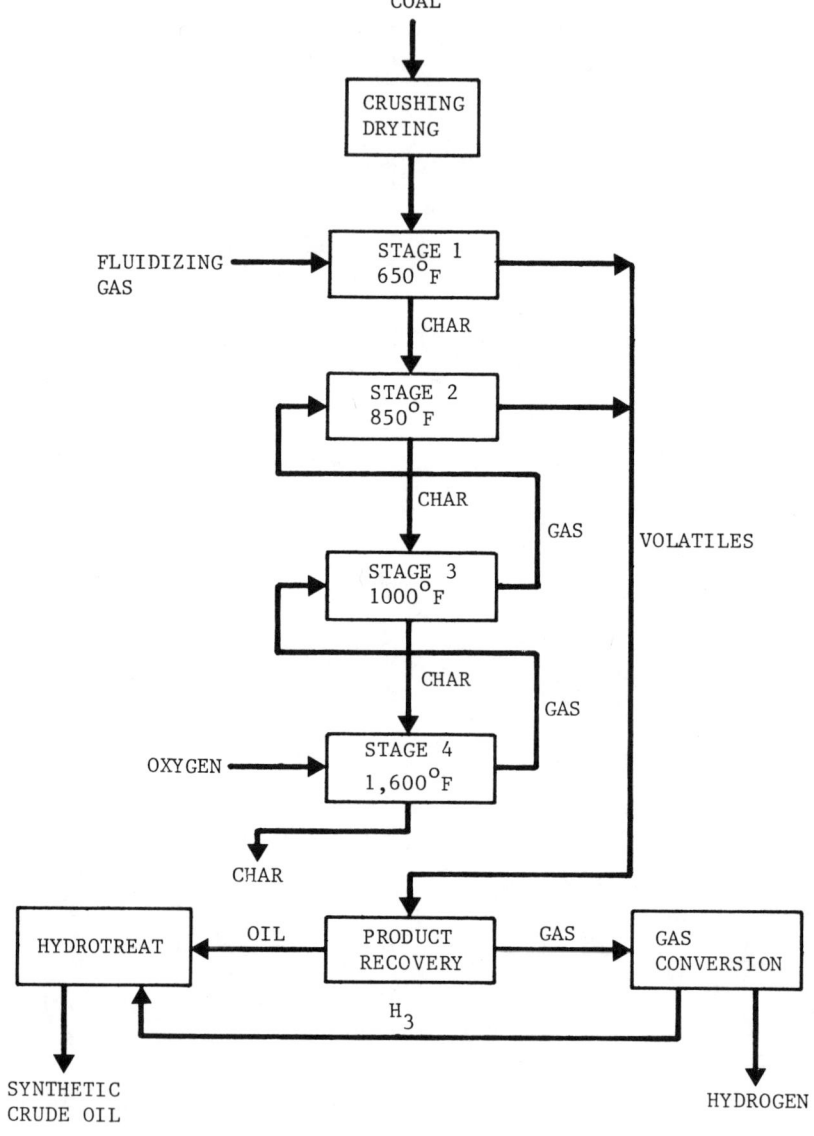

Figure 7-4. COED process flow diagram (source: Bloom and Wisdom, 1979).

The gasification section generates synthesis gas by reacting carbon with steam:

$$C + H_2O \rightarrow CO + H_2$$

MINOR GASIFICATION PROCESSES 251

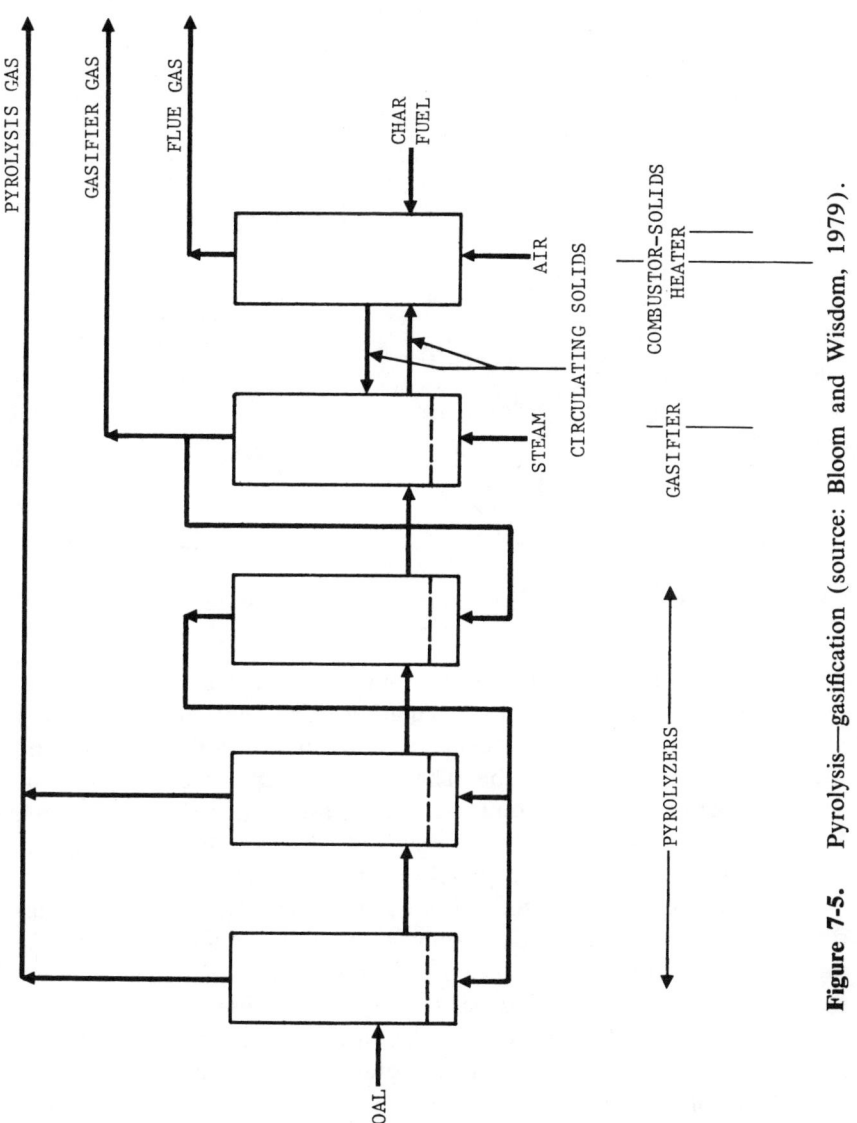

Figure 7-5. Pyrolysis—gasification (source: Bloom and Wisdom, 1979).

252 GASIFICATION

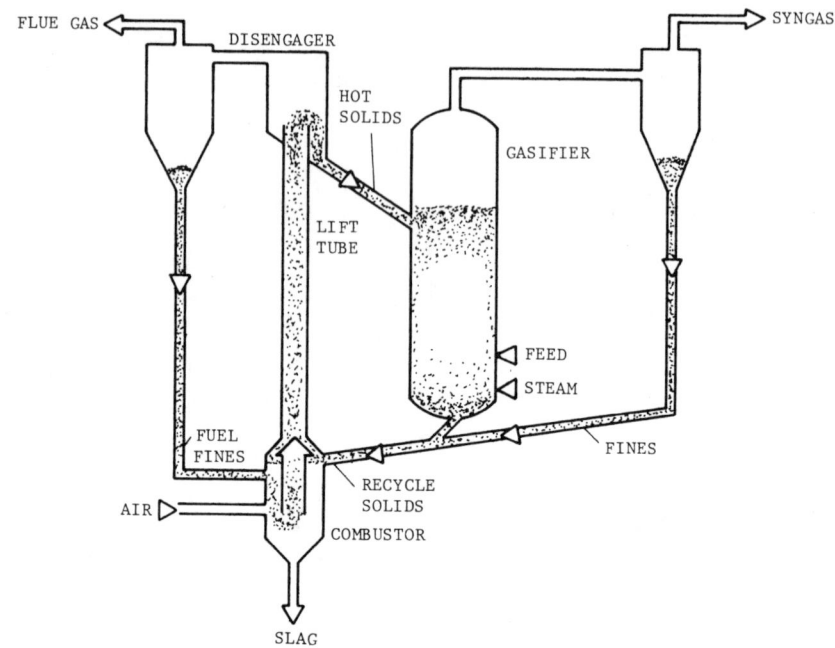

Figure 7-6. Gasification-combustion system.

The above reaction is endothermic, and needs heat to proceed. Heat for the reaction is provided by circulating chars through a combustor as shown in Figure 7-6. The heat exchange between the flue gas and the chars occurs in the lift tube. The chars are separated from the flue gas in the disengager and returned to the gasifier to react with steam. In this process, crushed coal is pressurized with synthesis gas from the process and sent to pyrolysis, where it is devolatilized by countercurrent-flowing synthesis gas. This section of the plant consists of several cylindrical steel vessels. The required number depends on the coal used; the more agglomerating the coal, the greater the number of stages required. For most coals, complete pyrolysis is accomplished in three stages.

The key feature of the process is that each stage is operated at a temperature below the coal softening temperature, eliminating the problems associated with the removal of molten ash and high carbon losses. The process avoids agglomeration by processing the coal in several stages, each operating at sequentially higher temperatures. The problem of coal agglomeration is also controlled by recycling some of the chars from the higher temperature stages to dilute the potentially agglomerating chars and to control the temperature. In the low-temperature stages,

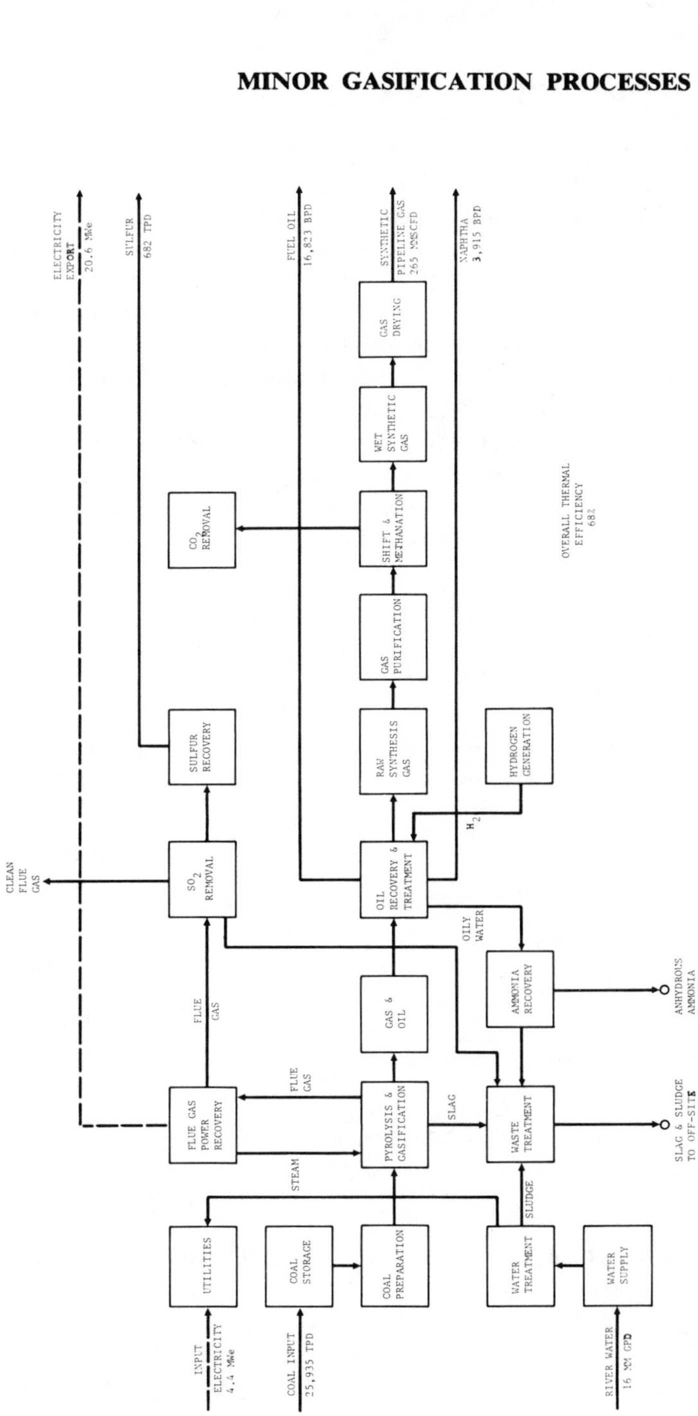

MINOR GASIFICATION PROCESSES 253

Figure 7-7. Illinois coal gasification group, COGAS commercial plant concept.

254 GASIFICATION

the oil partial pressure is controlled to accelerate the rate at which matter is removed from the chars.

The commercial plant schematically illustrated in Figure 7-7 will process Illinois No. 6 coal and water as feedstock and will produce substitute natural gas and fuel oil as principal products. Table 7-6 shows the composition of Illinois No. 6 coal. The coal and water input rates and the product and by-products are shown in Table 7-7. The compositions of SNG, fuel oil, naphtha and light hydrocarbons are presented in Tables 7-8 to 7-11, respectively.

The naphtha is produced as a result of upgrading the raw pyrolysis oil. The light hydrocarbons form during the pyrolysis and upgrading operation of the fuel oil; it is a mixture of alkanes and alkenes, with ethane and butane as the principal constituents.

Table 7-6. Properties of Illinois No. 6 Coal

Proximate Analysis (wt %)	
Moisture	12.08
Ash	13.27
Volatiles	30.80
Fixed Carbon	43.85
Total	100.00
Ultimate Analysis[a] (wt %)	
Carbon	76.55
Hydrogen	5.26
Oxygen	10.92
Nitrogen	1.11
Sulfur	5.95
Chlorine	0.21
Total	100.00

[a] Dry ash-free basis.

Table 7-7. Product and By-product

	Rate per Day
Input	
Illinois No. 6 Coal	25,935 tons (23,577 metric tons)
Water	16.52×10^6 gal (62.5×10^6 liters)
Output	
SNG	265×10^6 scf (7.5×10^6 normal^3)
Fuel Oil	16,823 bbl
Naphtha	3915 bbl
Sulfuric Acid	2178 tons (1980 metric tons)
Sulfur	682 tons (620 metric tons)
Anhydrous Ammonia	47.9 tons (43.5 metric tons)

MINOR GASIFICATION PROCESSES 255

Table 7-8. COGAS: Composition and HHV of Synthetic Pipeline Gas

Component	vol %
CO_2	0.14
CO	9 ppmv
H_2	0.57
H_2O	0.01
CH_4	94.00
Ar	0.04
N_2	5.24
	100.00
HHV, Btu/scf (J/normal m³)	950.9 (35.4)

Economic Analysis

The total capital requirements for the commercial plant are estimated to be $1,554,203,000 in mid-1978 dollars. The detailed breakdown is listed in Table 7-12.

Operating costs are estimated at $303,770,000 per year, as shown in Table 7-13. The biggest cost is associated with the acquisition of coal. For this estimate the cost of coal is $22.50/ton ($24.80/metric ton).

The price of the tailgate product gas is estimated for the first year of operation at $5.02/10⁶ Btu (4.76/GJ) under the utility financing method, and includes credits against operating costs from the sale of by-products. If all energy products from the plant are priced at an equivalent dollar value per million Btu of heating value (case 1), then the gas would be priced at $4.09/10⁶ Btu ($3.88/GJ). The basis for the first year analysis for utility-type financing is summarized in Table 7-14. The first year gas price calculations are shown in Table 7-15.

Table 7-9. COGAS: Fuel Oil Product Analyses

Composition, wt %	
Carbon	88.02
Hydrogen	10.65
Nitrogen	0.36
Sulfur	0.03
Oxygen	0.94
Total	100.00
Molecular Weight	226.00
°API	13.9
Pour Point °F (°C)	−30.9 (−34.9)
Flash Point, °F (°C)	175 (79.44)
Carbon residue on 10% bottoms (%)	0.0

256 GASIFICATION

Table 7-10. COGAS: Naphtha Product Analyses

Composition, wt %	
Carbon	86.62
Hydrogen	13.38
Nitrogen	<1 ppm
Sulfur	0.00
Oxygen	0.00
Total	100.00
Molecular Weight	93.60
°API	49.09
PONA Analyses	
Paraffins (vol %)	7.1
Olefins	0.0
Naphthenes (cycloparaffins)	
Monocycloparaffins	58.9
Dicycloparaffins	12.3
Tricycloparaffins	0.2
Aromatics	
Alkylbenzenes	19.8
Indans/tetralins	1.7
RONC	70.1

Table 7-11. COGAS: Composition of Light Hydrocarbons

Component	wt %
C_2H_4	7.5
C_2H_6	36.0
C_3H_6	9.7
C_3H_8	17.3
C_4H_{10}	29.5
Total	100.00

Table 7-12. COGAS: Commercial Plant Capital Requirement (mid-1978 dollars)

Capital Requirement	Costs (10^3 $)	Percentage
Plant Investment	1,275,800	82.2
Construction Loan Interest	186,683	12.0
Land	2,169	0.1
Startup	39,374	2.5
Administration	4,415	0.3
Working Capital	32,825	2.1
Initial Charge of Chemicals and Catalysts	12,937	0.8
Total Capital Requirement	1,554,203	100.0

Table 7-13. COGAS: Annual Operating Cost for a Commercial Plant

Classification	Annual Cost (10^3 $)
Coal	210,350
Labor	25,274
Chemicals and Catalysts	22,688
Insurance and Taxes	8,561
Repairs and Replacements	33.237
Utilities	1,158
Other Operating Supplies	2,502
Gross Operating Cost	303,770

Table 7-14. Bases for First-Year Economic Analysis (Utility Financing)

Total Capital	Fixed capital, construction loan interest, land, startup costs, administration, working capital, initial charge of catalysts and chemicals
Coal Price	Illinois, $22.50/ton ($24.80/metric ton)
Debt/Equity Ratio	75/25
Interest on Debt	9%
Depreciation	5% per yer, straight line
Return	15% on equity, after tax
Federal Income Tax	48%
Sulfuric Acid Credit	$52/ton (57.3 metric ton)

Table 7-15. First-Year Gas Price Calculations, Illinois No. 6 Seam Coal, Utility-Type Financing[a]

	$ million/yr			
	Case 1		Case 2	
	$/unit	Total	$/unit	Total
Fuel Oil, 18,415 bbl (2,927 kl)	24.78	150.59	15.40	93.59
Naphtha, 4,250 bbl (675.6 kl)	27.03	37.91	16.80	23.56
Light Hydrocarbons, 6,935 bbl (1,102.5 kl)	12.74	29.16	10.50	24.03
Sulfuric Acid, 2,266 tons (2,055 metric tons)	52	39.62	52	39.26
Ammonia, 52.4 tons (47.5 metric tons)	120	2.08	120	2.08
Total	4.09	337.83	5.02	414.31
Synthetic Pipeline Gas, 10^6 Btu[b]		259.00		182.52

[a]Total capital requirement = $1.5542 billion; coal: 28,330 ton/day (25,700 metric ton/day) at $22.50/ton ($24.80/metric ton) = $210.35 million; operating cost = $93.42 million; gross operating cost = $596.83 million.

[b]262.9×10^6 scf/day at 950.9 Btu/scf = 250×10^9 Btu/day.

Pilot-Plant Operations

The purposes of the pilot plant operations in Leatherhead, England, were:

1. to obtain additional hours of operation to show the technical feasibility of the COGAS process;
2. to obtain additional design information, including the confirmation or modification of earlier data, to be used in the design of a conceptual commercial plant and of a demonstration plant;
3. to operate the plant using chars from several coals; and
4. to test the utility of the proprietary COGAS material and energy balances in predicting the performance of the pilot plant.

Operations were started in February 1974. Three major tests were conducted in 1977–1978, during which 22 tests were completed. More than 1300 hours of gasification were logged, with the longest run testing 211 hours.

Other than process variables studies and model verification, the efforts were directed toward selecting the proper design of the proper mode of heat transfer to the gasifier and the combustor that would best fit the gasifier. The combustor selected was a vertical slagging cyclone-type combustor. This type is preferred because it offers the most advantage for heat transfer in fluidized conditions.

Two concepts addressing the problem of heat transfer to the gasifier were developed and tested. The inert heat carrier process illustrated in Figure 7-6 involves the use of inert pelleted refractory material which is heated in the lift tube right above the combustor by flue gases and recycled into the gasifier. This concept was abandoned in favor of the one presently in use. It was found that the use of pellets caused refractory lining failure and mechanical problems in the equipment.

Assessment

The COGAS process is in a design competition with the slagging Lurgi for a high-Btu demonstration plant. The winner is not expected to be chosen until after 1980. Since an integrated pilot plant has not been constructed, extrapolation of partial plant operations must be used to predict commercial size plant performance. Pyrolysis is considered a low-yield methodology for the production of coal-derived liquids.

BIBLIOGRAPHY

Bloom, R., Jr. "The Illinois Coal Gasification Group Project Incorporating the COGAS Process," paper presented at the American Gas Association Eighth Synthetic Pipeline Gas Symposium, Chicago, IL, October 1976.

Bloom, R., Jr., and L. I. Wisdom. "Chemical Feedstocks from Coal," paper presented at the meeting of the American Institute of Chemical Engineers, Houston, TX, April 1-6, 1979.

Kraayveld, H. J., and M. J. van der Burgt. "Technical and Economic Prospects of the Shell-Koppers Coal Gasification Process," paper presented at the 175th American Chemistry Society National Meeting, Industrial and Engineering Division, Anaheim, CA (1978).

INDEX

acid gas removal 56
acid removal system 57
air
    gasification for fuel gas manufacture 218
    pollution 109
    separation 170
Alberta tar sands 42
alkanolamines 127
ammonia recovery 173,205
analyses of feed coals 22
annual operating costs 196
applications of synthetic fuels 7
Ashland Oil Refinery 45
Atlantic Richfield Co. 14,43,45
atmospheric emissions 109,203
atmospheric gasifiers 236
Australian Gelliondale brown coal 43

Benfield system 128
Bergius process 38
bituminous coal 3,24
boiler efficiency 72
bottoms properties 26
British Gas Corporation 133,169
burner efficiency 219
by-product value 201

caking 164
calcium content 24
capital
    cost 4,76,115,182

investment 179
    requirements 185
carbon monoxide exhaust emission 220
Carter Oil Co. 14
catalyst 56
    summary 184
catalytic hydrogenation 38,42
chemical summary 186
Cities Service Research and Development Co. 42
Claus plant 128
Claus sulfur recovery 172
coal
    analysis 51,213
    and flux handling equipment 189
    concentration 68
    feed rate vs oxygen loading 158
    feed stock composition 73
    gasification 121
    liquefaction 26
    liquefaction residue evaluations 221
    mining 189
    operating conditions 140
    ranks 3
    storage areas 191
    swelling index 163
    tested for SRC-I 66
    transportation 190
    types 21
    utilization 3

Coal Research, Office of (OCR) 42
cobalt/molybdenum catalyst 38
COED process 248
    development unit 249
    flow diagram 250
COGAS
    annual operating cost 257
    commercial plant capital requirement 256
    commercial plant concept 253
    composition 255
    composition of light hydrocarbons 256
    fuel oil product analyses 255
    naphtha product analyses 256
    process 248
combined cycle plant 236, 237
combustion 129
    chamber parameters 219
Combustion Engineering (CE)
    two-stage gasifier 122,224
    entrained bed coal gasification process 224,226
combustion test results 72
commercial H-coal liquefaction plant 50
commercial plant design 73
commercial size gasification plant 135
commercial utilization 8
component scale-up 170
conceptual commercial plant designs 29
Conoco Oil Company 169
Consolidation Coal Co. 99
Continental Oil Co. 45
cooling water system 173,207
cost 20
    estimates 9
    of construction 73
    of gasification product 247
    of upgrading coal liquids 113

design of the 250-ton/day pilot plant 17

development programs 18
devolatilization 129
    reaction rates 130
diesel oil 97
distribution torque 163
donor solvent process 20
dry-ash Lurgi gasifier 155
drying 129
    and compressing 172,205

ebullating bed reactor 37,38,41
ECLP Test Program 18,19
economic 7
economic analysis 58,89,91,113, 175,247,255
    liquefaction process 6
    of coal liquefaction 114
economic evaluation 234
economic parameters 4
economy of scale 7
EDS
    coal naphtha 25
    commercial plant 30
    math modeling 26
    process alternatives 28
    products 25
electric power generation 8
Electric Power Research Institute 14,45
emissions 219
entrainment gasifier 230
environmental analysis 223,238
enviromental evaluation 188
environmental problems 105
equipment sizes for H-coal 46
Estosolvan 128
ethane-propane gas into ethylene 76
European utilization 7
Exxon Donor Solvent (EDS) 13,14
    Commercial Plan Study Design 31,32,33
    See also EDS

feed fines concentration 163

feedstock 3,112
filtration flowsheet 64
Fischer-Tropsch commercial plant
    mass balance 98
Fischer-Tropsch method 82,87,
    93,96
fixed-bed pilot plant 82
flare and incineration 174,208
flare gas analysis 161,162
flexicoking 13,26,28
    process 21
flowsheet
    for the SRC-II process 67
    for Wilsonville SRC pilot plant
      62
    of Texaco process 214
flue gas desulfurization 206
fluid-bed combustor 101
fluid-bed reactors 95
Fluor solvent 128

gas cleanup systems 123
gas cooling 172,201
gas cost variation 189
    with capital investment 191,193
    with coal 190
    with coal cost 198
    with debt interest rate 199
    with operating cost 192
    with plant yield 202
gasification 129,144,170
    characteristics of 124
    combustion system 252
    processes 121,133
    summary of 122
gasifier
    combustor 230
    material and energy balances
      232
    operating condition 138
    operating parameters 212
    performance 213
gas liquor separation 173,205
gasoline blending feedstock 25

H-coal

commercial plant design 52
complex block flow diagram 54
ebullated-bed reactor 41
economics 58
effluent 55
    pilot-plant 47
plant design 55
process 37,39
process history 40
process yields 52
reactor 53
syncrude vacuum tower bottoms
    222
health problems 105
heat and material balance 176
heat recovery 150
heavy fuel oil 97
high-Btu gas manufacture 144
high-pressure product gases 34
high-sulfur coal 37
Hydrocarbon Research, Inc. 37
hydrocarbon synthesis 95
hydrocracking 63
hydrogenated recycle process 16
hydrogenation 38,63
    feedstocks 27
hydrogen donor solvent 26
hydrotreating 25,34,111
    cost of 111

Illinois coal gasification group 253
Illinois #6 coal 179
    base case design 29
Institute Francais du Petrole 128
instrument air systems 173,208
investment costs 90
investment expenditure schedule 31

Japan Coal Liquefaction
    Development Company 14

kinetic equations 130
Koppers-Totzch process 122

large engineering projects 9
lignite 3

264 COAL LIQUEFACTION AND GASIFICATION

linear programming model 28
liquefaction
  bottoms 109
  direct, American processes 11
  pilot units 20
  process data summary 5,13
  process parameters 4
  process problems 107
  process steps 110
  product market studies 24
  products 111
  reactor system 32
  residue evaluation 220
liquid product yields 23
low-pressure separator 33
Lurgi 133
  dry-ash process 168
  pipeline gas plant 134
  process 122
  low-pressure methanol synthesis process 85

maintenance and operating expense 200
major gasification process 133
material balance 167
  gas cooling 150
  gasification 141
  methanation 155
  shift conversion 145
  sulfur recovery 157
materials preparation 169
mathematical modeling of gasifier performance 128
mechanical equipment 107
medium-Btu gas 174
metallurgical requirements 230
methanation 85,172,204
methanol
  to dimethyl ether 81
  to gasoline process 84
  to hydrocarbons 81
mineral matter 33
minor gasification processes 243
minor liquefaction processes 79

Mobil
  commercial plant design 85
  gasoline synthesis 79
  methanol conversion process 93
  methanol to gasoline 80,88
  Oil Corporation 45
modeling EDS process 26
molecular sieve treaters 56
monoethanolamine 34,86

naphtha 254
  hydrotreater 56,57

operating
  conditions 4
  cost 77,90,184
  cost data 239
  expenses 32
operation of pilot plant 34
oxidation of bottoms 59

performance of internal components 164
petroleum hydrotreater 82
phenol extraction 173,205
Phillips Petroleum Co. 14
pilot plant 31
  flowsheet 217
  gasifier 164
  operation 35,71,151,211,258
  yields 50
pipeline gas 92
Pittsburgh #8 coals 29,169,179
plant
  cost summary 180
  design 8,55
  economics 6
plug flow liquefaction reactors 16
plug flow moving bed coal gasifier 128
pollution potential 204
Powhattan Coal Case 74
preheater design 32
price structure 178
process
  comparisons 116

costs   116
economics   102
selection   11
product
   fractionating   97
   gas composition   161,162,237
   value factors   115
   yields in SRC-I and SRC-II modes   70
profitability   4
program costs   175
properties of Illinois #6 coal   254
Purisol   128
pyrolysis-gasification   251
pyrolyzing process   248

railroad network   3
raw material summary   184
recovery of liquid products   21
Rectisol   128,150,172,204
recycle gas cleanup   34
recycle solvents   16
reformer naphtha   57
Roland seam coal   131

scale-up   11
scrubbing solution   34
Selexol   128
sensitivity analysis   185
Shell-Koppers   243,244
   coal gasification coal feed analyses   245
   coal gasification dry synthesis gas composition   245
   gasifier operating conditions   245
   processes   122
Shell Oil Company   45
shift conversion   145,149,171,199
shipping   174
   and receiving facilities   208
slag
   composition and steam/oxygen ratio   160
   handling   173,207
   tapping   160
slagging Lurgi   133,168

SNG
   cost of production   197
   product gas cost   183
   properties   136
   selling price   194
soil erosion   193
solid waste disposal   108
solvent hydrogenation   20
Spencer Chemical Co.   67
solvent refined coal (SRC)   60
solvent-to-coal ratio   35
spent donor solvent   34
SRC-II commercial plant   75
SRC-II process description   61,63,66
Standard Oil Company   45
steady-state differential equations   130
steam
   generation   173,206
   plant with atmospheric gasifiers   235
   reforming   29
stream analysis
   methanation   156
   Rectisol   152
   shift conversion   146
Stretford process   127,225
sulfur   127,227
   content   24
   recovery   172,205
Sun Oil Company   45
support facilities   174
synthetic natural gas
   See SNG
system analysis: gasification   142

tankage   174
   and fuel oil system   208
tar recycle   161,162
Texaco coal gasification process   209
Texaco gasifier   210,216
Texas lignite   29
thermal efficiency   29,246
   of commercial H-coal plant   53

turbine fuels   113
Turbo Power and Marine Systems, Inc.   215
types of coal   21

ultimate analysis, Kentucky coals   70
Union Carbide amine guard process   85,86
unit costs   92
United Aircraft Corp.   215
utility financing economics   193,195,196

volatile material   24

waste streams   105

wastewater treatment   107,174
water
  pollution   191
  runoff   193
  treatment   173,206
Wellman-Lord unit   206
Westfield Development Centre   169
Wyodak coal   35,44
Wyoming
  coal   15
  strip-mined coal   92
  subbituminous coal   87

zinc chloride catalyst   99,100
zinc halide hydrocracking process   99